Fruit and Vegetables

Ian MacDonald and John Low

Evans

Evans Brothers Limited

Evans Brothers Limited
2A Portman Mansions
Chiltern Street
London W1M 1LE

Evans Brothers (Nigeria Publishers) Limited
PMB 5164, Jericho Road
Ibadan

Evans Brothers (Kenya) Limited
P.O. Box 44536
Nairobi

First published 1984

Illustrated by Annette Olney

Typeset in 10 on 11 point Times Roman
Printed by BAS Printers Limited
Over Wallop, Stockbridge,
Hampshire

ISBN 0 237 50790 9

Evans Books for Rural Development

TROPICAL FIELD CROPS
FRUIT AND VEGETABLES
COMMUNICATION SKILLS FOR RURAL DEVELOPMENT

Contents

Introduction

This book has been written for agricultural advisers, teachers and farmers who require a reference book that contains basic information about growing crops. There are already a number of agricultural science books available, but these generally do not go into detail concerning essential farming practices. *Fruit and Vegetables* attempts to fill this need.

This book deals with those crops that are grown in a vegetable garden or small orchard. Crops that are grown on a large scale, such as coffee, tea, cotton, rubber, sisal and groundnuts, are dealt with in another book in this series, *Tropical Field Crops*. The recommendations given are designed to be applicable throughout the tropics, but since weather conditions and soil will vary considerably, advice should always be sought from local experts, particularly with regard to the best varieties to plant, fertilizer requirements, and the control of pests and diseases.

Technical terms that have been used are explained in a glossary at the back of the book. Also included at the back of the book are a list of chemicals and their various trade names to make it easier for farmers to purchase the correct chemicals. The production figures given are for tropical countries only, and are taken from the FAO Yearbook for 1979.

We are grateful to Dr Dick Hood for advice on chemicals, and to a great many other colleagues in Africa and Asia for their generous help.

Ian MacDonald
John Low

Section 1
Guidelines on Fruit and Vegetable Production

Soil

The soil for vegetable growing should, if possible, be a sandy loam soil. It should be well-drained so that it does not become water-logged.

The yield and quality of crop a farmer can grow depends on the soil, and in particular on the amounts of *nutrients* (food for the plants) to be found in the soil. Soil which is planted regularly often contains too little food for the plants, and fertilizer and manure therefore need to be added. But fertilizer is expensive, so farmers will not want to use more than is necessary.

A soil-testing service provides the best way of finding out how much fertilizer needs to be added. In many countries, the Department, or Ministry of Agriculture provides a soil-testing service which is free, or very cheap, to small and medium-sized farms. Instruction on how and when to take a soil sample for testing will be available from the soil-testing station. This service can often save the farmer money, by telling him how much fertilizer he needs to apply. The soil-testing station can also check that the soil provides the correct conditions for a particular crop.

Soil pH

An important factor in soil is whether it is *acid* or *alkaline*. A soil-testing station can tell the farmer which type of soil he has. The result of the test may be given in the form of a *pH value*. If this figure is less than 7·0 it means that the soil is acid, but if the figure is greater than 7·0 the soil is alkaline. If the figure is very close to 7·0 then the soil is neutral. Most vegetables do not like very acid soils, and the best pH range for vegetable growing is 6·0–7·5. Brassicas, onions and lettuce are particularly sensitive to acid soils and will not grow well where the Ph is below 7·0.

Water supply

A permanent water supply near the vegetable beds is necessary. If the supply is from a stream, then dig a well a few metres away from the bank so that nematode (eelworm) infection is prevented.

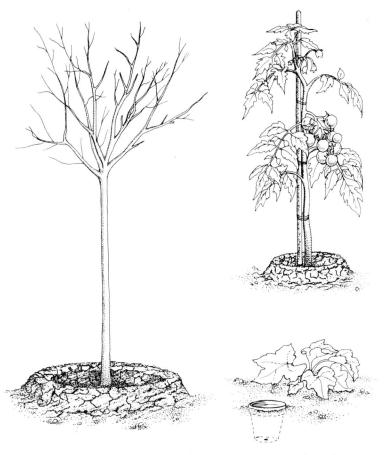

Figure 1   Make an irrigation ring or pour water into a buried pot

To make more effective use of the water you can build up earth to form an irrigation ring around each plant, as shown in Figure 1. This is particularly useful for fruit trees. Another way of making watering more efficient is to bury a pot or tin in the soil beside each plant and water into the pot, as shown in Figure 1. The pot or tin should have a hole in the bottom. This is especially recommended for tomatoes and cucurbits. As well as saving water, it stops the plant stems from getting wet during watering, and this can help to protect them from stem-rot diseases.

Wind protection

Wind can cause considerable damage to vegetables and fruit trees. It may break off leaves and branches, and this helps diseases to infect the plant. Wind may also dry out the topsoil and the plants then suffer from lack of moisture (see Figure 2). To protect against wind, strong fences should be made or tall boundary crops planted.

Figure 2   Wind damages vegetables so build a fence

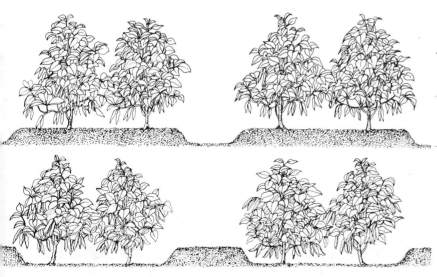

Figure 3   In the wet season raised beds improve soil drainage, but for irrigation sunken beds are best

Land preparation

In very wet areas, vegetable beds should be raised to prevent water-logging, as shown in Figure 3. In dry areas, however, sunken beds should be made so that as much water as possible is retained.

The soil should be dug or ploughed as soon as possible after the previous crop has been harvested. If crop residues are dug or ploughed in, this should be done while there is still some moisture in the soil, so that the crop residues can rot down. Before planting, the soil should be hoed to break up large clods of earth, or disced or harrowed if mechanical cultivation is being used.

Cultivating slopes

Great care should be taken in cultivating slopes, as soil erosion can easily occur. Rows of vegetable plants or fruit trees should always follow the contour, as shown in Figure 4. This reduces the risk of erosion.

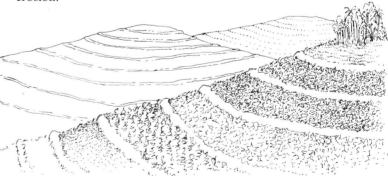

Figure 4   Plant in rows along the contour to prevent erosion

It may be possible to terrace sloping land, but this will only be worthwhile if highly profitable crops are then grown on the terraces. The essential features of a terrace are shown in Figure 5 on page 4.

Above the highest part of the terrace an extra-large drain should be dug to carry away rainwater during storms. This storm channel should continue at a very gentle slope to the bottom of the hill, so that storm water can run off without causing soil erosion. Alterna-

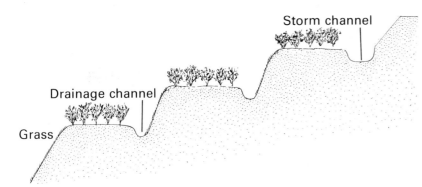

Figure 5 *Slope terraces so that rain water flows to the drainage channel*

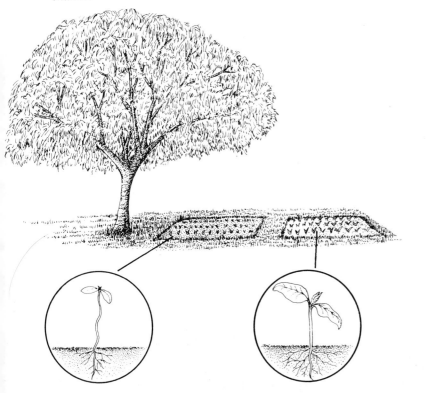

Figure 6 Too much shade causes seedlings to grow sappy and weak

tively, it should run off into an existing water channel. Put dividers into the storm channel at 15 metre intervals. The dividers should be about half as deep as the channel. These help to slow down the flow of water, reducing soil erosion and allowing some of the water to soak into the soil.

Choosing a nursery site

When young, many plants require extra care, so it is best to grow them in a special area, known as a nursery. This also saves space in the main vegetable beds, cuts down the amount of hoeing that needs to be done early in the growing season, and makes watering of the seedlings easier.

Sites for vegetable nurseries should be chosen carefully. The site should not be on a slope (unless terraced) because soil erosion could become a problem. Nurseries should not be sited where they are heavily shaded by trees or buildings, as seedlings grow poorly under these conditions, but a lightly shaded area may be best in very hot regions.

Tree seedlings are best grown in pots, rather than in the ground, so that they can be transplanted without any damage to the roots. Growing them in pots also saves space in the nursery.

Seeds

It is important to plant good seed that will germinate well and produce vigorous seedlings. Many plant diseases are carried in the seeds (they are said to be 'seed-borne') and if diseased seed is used the plants that grow from it will suffer from the disease.

Commercial seed from a good seed merchant is generally certified free from disease. This is often referred to as 'certified seed', and although it may be more expensive it is worth buying because it will always give healthy seedlings.

Another advantage of using commercial seed is that it will always produce plants that are 'true to type' (i.e. they have the characteristics expected of their variety).

For some crops, commercial seed may be unavailable, so that the farmer has to use his own seed. Only the best seed from healthy plants should be used. It should not be harvested until it is fully mature, and after harvesting it should be left to dry for several days before being stored in tins with good lids. The tins should be kept in a cool, dry place. The seed must be used in the following season; if it is kept for any longer it will not germinate well.

Some plants, such as peppers, often give poor results when grown from non-commercial seed, because the seeds are the result of cross-pollination with another variety. Never collect seed from an F_1 hybrid variety (of maize or tomatoes, for example). The plants that grow from this seed will be of very poor quality.

Seed dressings

Seed is usually mixed with chemicals ('dressed') before planting to protect it against insects and fungal diseases. Most commercial seed is already dressed, ready for planting.

If non-commercial seed is used, it should be dressed before it is planted. There are several seed-dressing mixtures available which contain both an insecticide and a fungicide. Alternatively, a single chemical compound may be used; some are recommended on p. 133.

If a particular pest or disease is prevalent, seed dressing with a certain chemical compound may be recommended in the **Pests and Diseases** section. Use the one recommended in place of the usual seed dressing.

WARNING Seed dressing compounds are all very poisonous. They are not removed from the seed by washing and cooking, so never eat dressed seed. Remember that seed bought from a seed merchant will usually be dressed, so it must not be used for food. After handling dressed seed, always wash your hands with soap and water.

Sowing

Water the nursery beds thoroughly the day before sowing, as heavy watering after sowing may wash small seeds away. If you are just growing vegetables for your family, only sow a small amount of seed at any one time. Provided the rainy season is long enough, or irrigation is used, a new batch of seed can be sown every few weeks. This will give you a regular supply of fresh vegetables.

Seed should not be sown too deeply. Small seeds should be sown just below the surface. Larger seeds should be between 2 cm and 8 cm deep, depending on the size.

Figure 7   Rake the soil, mark out drills, sow the seed carefully, lightly cover with earth and firm down

Mulching

In dry weather, mulching is very important, in both the nursery and the vegetable beds, as it helps the soil retain moisture and stops it from getting too hot. Paths should be mulched as well, to prevent the edges of the beds from drying out (see Figure 8).

Chopped grass is useful as a mulch, but do not use grass that has gone to seed unless you remove the seed heads first. In the nursery, thin the mulch as soon as the seedlings appear. Mulch should not touch the stems of the plants, as this makes it easy for pests to attack them.

Figure 8   Mulch prevents the soil drying out

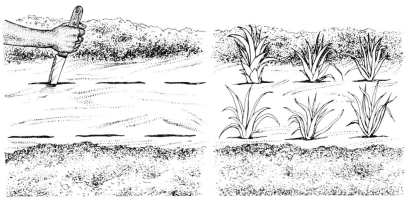

Figure 9   Black polythene keeps down weeds and acts as a mulch

Black polythene can also be used as a mulch in vegetable beds, but not in the nursery. Before transplanting, spread a sheet of polythene on the ground and cover the edges with soil, as shown in Figure 9. Make cuts in the plastic where you will want to put the seedlings, and plant the seedlings through these holes.

Providing shade

Young seedlings may require some extra shade to protect them from the sun. A shade can be built as shown in Figure 10. The shade should allow some light through, and it should be thinned as the plants grow.

Figure 10   Build a shade to protect seedlings from the hot sun

Figure 11   Thinning out should be done before the seedlings grow too tall

Thinning out

Once the seedlings are a few centimetres tall, you may have to pull some of them out and throw them away to give the others more room (see Figure 11). Pull out the small, and unhealthy-looking seedlings. This process is called 'thinning out'. Seedlings should be thinned out as soon as possible in order to obtain strong plants. If they are left for too long, they will become very tall and weak.

Hardening off seedlings

Seedlings are ready to transplant when they have four true leaves (see Figure 12). About a week before transplanting, let the seedlings have only a little mulch or shade, and give them less water than before. This will 'harden' them so that they can survive when transplanted. Weak, thin seedlings may not survive transplanting, and even if they do they will not grow well.

Figure 12   The seedling on the right is ready to transplant

Transplanting

Hardened-off seedlings can be planted out into their final growing places. Transplant in the evening or on a cloudy day so that the plants are not immediately exposed to the hot sun.

When transplanting, take care to disturb the roots as little as possible. Once the seedlings have been taken from the nursery, do not leave them lying around for long. If they have to be left for a short while, dampen their roots. Follow the spacing recommended and plant them firmly in the soil. Water the seedlings in, if possible. The seedlings may wilt at first, but they should recover after a day or so.

Manure and compost

Most vegetables grow best on soils that are rich in organic matter, so add as much animal manure or compost as you can to the beds before planting. A rate of 1 kg of manure or compost for every square metre of land is ideal (1 kg is about as much as you can hold in your two hands, and a metre is about equal to a man's stride.) The manure of compost should be mixed in well with the soil, so that it does not come into direct contact with the roots of the crop.

Manure and compost must be properly rotted so that any weed seeds present have germinated and been killed before it is used. Manure needs to rot down for about six months before use. Compost takes about the same time, but this varies, depending on the climate; it is ready for use when it is dark in colour and crumbly. Use the oldest compost heap first.

Making compost

Compost is made from weeds, leaves, crop residues and vegetable waste from the kitchen. Do not add bones or meat scraps which will attract rats, dogs and wild animals. Compost can be made quite easily in a heap, or in a simple container, such as those shown in Figure 13. The leaves and vegetable waste are put into the container to a

Figure 13   Some simple containers for compost

Figure 14   Build up layers of plant and starter material

7

depth of about 20 cm and then a layer of starter material is added. These starter materials are things such as manure or rotting fruit and vegetables which contain the bacteria needed to make the compost. The heap is built up with alternate layers, as shown in Figure 14. In dry weather the compost heap should be watered, to allow the bacteria to grow. After 2–3 months the heap should be mixed up well.

Fertilizer

The three main fertilizers are nitrogen, phosphate and potash. These are sometimes shown as N, P and K. Below is some information about these fertilizers.

Nitrogen

This is required to encourage rapid growth and good green colour. Too much nitrogen causes weak, soft growth and poor fruiting. Nitrogen can be easily washed out of the soil. Commercially it is sold as:
ASN ammonium sulphate nitrate
CAN calcium ammonium nitrate
SA sulphate of ammonia: It makes soil more acid; use it on alkaline soils only.
Urea made from ammonia and carbon dioxide: Should be worked into the soil immediately after application. Has approximately twice the nitrogen content of the other fertilizers.

Phosphate

Phosphate encourages good root growth and early maturity. It is sold commercially as:
DSP double (or triple) superphosphate: Makes soil less acid.
SSP single superphosphate: Makes soil less acid; half the strength of DSP.
DAP diammonium phosphate: This also contains nitrogen.

Potash

Potash influences the intake of plant foods and crop yields. If deficient, usually a whole area will be deficient, and this will be known by local agricultural advisers. Potash can be supplied as muriate of potash, or in a compound fertilizer.

Compound fertilizers

Compound fertilizers are mixtures of different fertilizers. They contain nitrogen (N) and phosphates (P), and generally some potash (K) as well, and they are therefore often called 'NPK fertilizers'. However, the proportions of these three different nutrients vary from one compound fertilizer to another. The composition of a compound fertilizer is described by a series of three numbers, such as $25-25-5$. The first number always refers to the nitrogen content, the second number refers to the phosphate content, and the third number the potash content. So in this case, the compound fertilizer contains 25% nitrogen, 25% phosphate and 5% potash.

Compound fertilizers are recommended for certain crops, such as maize, where they are useful in giving the crop the correct nutrient balance. For crops of this kind, compound fertilizers are often the most economical way of supplying nutrients.

Minor nutrients

Apart from nitrogen, phosphates and potash, crops need a number of other nutrients, but they need them in fairly small quantities, and they vary in their requirements for them.

Where deficiencies of minor nutrients are likely to affect a crop, this is mentioned in the second section of the book. A soil-testing station can advise you if your soil is suitable for a particular crop.

These minor nutrients may be supplied to the soil in several different ways. One way is to apply a compound fertilizer. For example, the compound fertilizer called $8-21-24+2Mg$ supplies magnesium, Mg being the symbol for magnesium. (The symbols for some other minor nutrients which you may come across are as follows: iron–Fe, copper–Cu, zinc–Zn, boron–B, calcium–Ca, sulphur–S, manganese–Mn, sodium–N, chlorine–Cl, molybdenum–Mo.)
Another way in which these minor nutrients may be supplied is in a leaf spray (also known as a foliar spray). This may be sprayed alone, or added to a pesticide spray to reduce the amount of work involved.

A third method of supplying minor nutrients is by applying a compound or a chelate of the mineral concerned, scattered on the soil around the plant. A chelate is a hard, granular material which slowly releases the nutrient into the soil. Chelates are often used for fruit trees.

Measuring out fertilizer

The fertilizer recommendations in this book are written in three different ways:

1. Kilos per hectare (kg/ha) is generally the most accurate, but will only be useful if mechanical fertilizer applicators are used.

2. For the farmer who spreads his fertilizer by hand, amounts are given for a certain number of paces (a tall person's pace is about a metre in length) along a row of plants that one kilo of fertilizer should cover.

3. The third method is to measure out the fertilizer with a spoon. This is the best way when a small quantity is required. The 'large spoon' referred to in this book is the sort of spoon that is used for eating. The size is shown in Figure 15. The 'small spoon' referred to in this book is the sort of spoon used for stirring tea, and its size is shown in Figure 16. When the large spoon is heaped up with fertilizer it gives about 20 gm. When the small spoon is heaped up with fertilizer it gives about 6–7 gm.

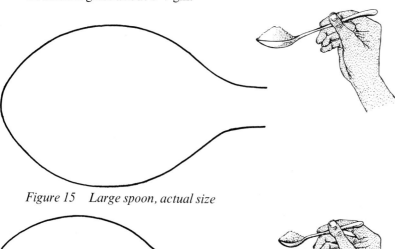

Figure 15   Large spoon, actual size

Figure 16   Small spoon, actual size

Crop rotation

Some pests and diseases can survive in the soil and attack the next crop that is planted there. If the same crop is grown on a piece of land year after year, the diseases which affect that crop may build up to a very high level in the soil. The crop will be severely attacked and its yield greatly reduced.

Changing to another crop, which is not attacked by the same pests and diseases, will prevent this build-up, and a different crop should therefore be grown on each field or vegetable bed every season. The new crop should belong to a different family from the previous crop, because certain diseases tend to attack plants from the same families.

Crop rotation systems divide plants into groups according to which family they belong to. The land is divided into the same number of plots as there are groups, and each plot is planted with crops from a different group in each season.

For the first season plot 1 would be used only for crops in group 1, plot 2 for crops in group 2, plot 3 for crops in group 3 and so on. For the next season the groups are all changed round, so that plot 1 is used for group 2 crops, plot 2 for group 3 crops, plot 3 for group 4 crops and so on. As well as helping to control pests and diseases, crop rotation has a beneficial effect on soil fertility, because different crops take different kinds of food from the soil and from different depths. If one crop is grown continuously the soil may become deficient in certain nutrients. On the other hand, crops such as peas and beans return some nitrogen to the soil, which can benefit the next crop.

A five-group rotation using the groups shown below, is recommended:

Group 1 (Brassicas)	Group 2 (Solanacs *etc*)	Group 3 (Cucurbits)	Group 4 (Roots and	Group 5 (Pulses)
Cabbages	Brinjals	Squash	Carrots	Beans (all types)
Kale	Peppers	Marrows	Celery	Peas (all types)
Brussels	Potatoes	Pumpkins	Onions	Lentils
sprouts	Tobacco	Cucumbers	Garlic	Grams
Cauliflowers	Tomatoes	Gherkins	Leeks	Lablab
Turnips	Cape goose-	Melons		Groundnuts
Radishes	berry	Gourds &		
	Okra	Calabash		

If a five-group rotation system cannot be practised, then a three-group rotation should be used instead. However the three-group system is less effective in controlling pests and disease, and a five group rotation should always be used if possible. A three-group system is shown below:

Group 1	Group 2	Group 3
(Cucurbits and Pulses)	(Solanacs, Roots and Bulbs)	(Brassicas)
Groups 3 and 5 in the table on p. 9	Groups 2 and 4 in the table on p. 9	Group 1 in the table on p. 9

A pest that affects a number of different crops is the nematode or eelworm, a very small worm that lives in the soil and attacks plant roots. This pest may build up to a high level in the soil, even though crop rotation is practised. If nematodes become a problem, plant grass or cereals on the infested land for a few years, as nematodes do not affect these crops.

Intercropping

Intercropping means growing more than one crop at a time in the same field. Until recently, little research was done on intercropping, although it is a very common way for farmers in the tropics to grow their crops.

Nowadays it is accepted that intercropping can sometimes be of benefit to the small farmer.

The main benefits of intercropping are:

1. The total farm output is generally raised, and does not vary so much from year to year.
2. A greater variety of food is produced, and there is less risk to the farmer than if he relies on just one crop.
3. Less weeding needs to be done.
4. The damage done by pests and diseases may, in some cases, be reduced.

The disadvantages of intercropping are:

1. Weeding cannot be mechanized, and has to be done more carefully.
2. There are very few herbicides that can be used on mixed crops.

Pests and diseases

Most crops are attacked by various insects, and by other pests, such as nematodes, birds and rodents. Crops can also be damaged by a number of different diseases. Some crops, such as citrus fruits, are affected by a great many pests and diseases, while other crops, for example leeks, have very few disease problems.

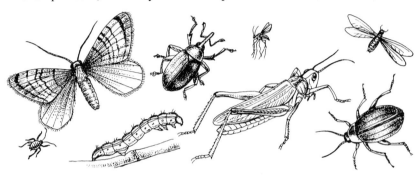

Figure 17   Some insects which may damage crops

Good management will greatly reduce the damage caused by pests and diseases. The four basic management practices are as follows:

1. Field hygiene
   Immediately after harvesting, dig in, plough in or burn all weeds and stubble. These crop residues often harbour pests and diseases. Diseased plants should generally be burned.
2. Certified seed
   As explained on p. 4, certified seed is free of disease. Using certified seed will greatly reduce crop losses due to disease.
3. Promotion of vigorous plant growth
   Use manure, compost and fertilizers to improve growth, because vigorous plants are more resistant to disease.
4. Crop rotation
   A crop rotation system will help to prevent carry-over of disease.
5. Resistant or tolerant varieties
   Some varieties are resistant to attack by certain pests or diseases. These resistant varieties should be grown when a pest or disease is particularly serious in your area. There are also varieties that are tolerant to certain pests or diseases; these are attacked, but they do not suffer as much damage as other varieties.

If pests and diseases become a problem despite these management techniques, then treatment with chemicals may be the only way to save the crop. The use of chemicals is described on pp. 11–13. Specific instructions about the treatment of pests and diseases are given under each crop in the second section of the book.

Different types of plant disease

There are four different types of plant disease:
1. Fungal diseases
2. Bacterial diseases
3. Viral diseases
4. Physiological diseases

Fungal diseases

Most fungal diseases do not destroy the crop completely, but they do reduce the yield. Fungal diseases can spread from one plant to another in the air, in the soil, on the seeds, by insects or by water splashing the leaves. The spread of fungal diseases can be reduced by good management as described on p. 12. These diseases can also be controlled by the use of chemicals.

Bacterial diseases

Bacterial diseases can destroy a crop completely. They cannot be controlled by chemicals (except by using very expensive methods which are not worthwhile for small growers). They are usually carried in the soil or in seeds, so the best way to control them is by planting certified seed and following the other management practices described on p. 4.

Viral diseases

Viral diseases generally cause mottling, streaking or curling of the leaves, and stunt the plant's growth. They do not usually destroy the crop but they can reduce the yield considerably, especially in fruit crops. Viral diseases cannot be controlled by any chemicals. They are carried from one plant to another by several means:
1. By sucking insects such as Aphids and White Fly

2. By people touching the plants
3. By pruning knives and other garden tools
4. By taking cuttings from infected plants

The only way to control viral diseases is to burn all infected plants and to kill Aphids and White fly so that they cannot spread the diseases. In some cases it is a good idea to disinfect the hands and garden tools, to avoid carrying the disease from one plant to the next, for example, when pruning fruit trees or vines.

Physiological diseases

Physiological diseases are different from other types of disease. Fungal, bacterial and viral diseases are all caused by tiny organisms (fungi, bacteria or viruses) which enter the plant from outside, but physiological diseases are *not* caused by an organism infecting the plant. They usually appear because the soil does not have the correct balance of nutrients for the crop (see p. 8) or because the plants have been given the wrong amount of water.

Using chemicals for pests and diseases

In many cases it will be necessary to use chemicals to control pests and diseases. However, chemicals should only be used when absolutely necessary. They are expensive, so if the crop is not a particularly valuable one it may not be economic to use them. They are also dangerous, if misused, and you should read pp. 12–13 carefully before spraying. All chemicals must be used at the recommended strength to be effective. Spraying less (or more) than the required amount will simply be a waste of money.

Pests

The main types of pests controlled by chemicals are insects and nematodes (eelworms). Insects are controlled by using *insecticides* and nematodes by using *nematicides*. (The word *pesticide* is also sometimes used for these chemicals. It includes any chemical that kills pests.)

Spray insects only when the attack is severe. There is no point in spraying just to kill a few insects. Large insects can often be removed and destroyed by hand instead of spraying. Before you decide to spray, make sure the insects are really harmful to the crop – not all insects are.

For most insect pests you should spray once, then check the crop the next day to see if the pest has been killed. If it has not you should seek expert advice.

Diseases

Viral, bacterial and physiological diseases cannot be controlled by chemical spraying.

The chemicals used to control fungal diseases are known as fungicides, and they are applied by spraying or as a seed dressing. However, with many fungal diseases the disease cannot be controlled once it has got a good hold on the plant. In these cases preventive spraying is advisable if the crop is a valuable one.

When preventive spraying is recommended it is absolutely essential to begin spraying before there are any signs of the disease and continue at regular intervals throughout the growing season. To control most fungal diseases the crops need to be sprayed with a fungicide every two weeks. Spraying the crops less often than this is generally not worthwhile. In hot, damp weather there is more risk of the disease developing and spraying should be carried out even more frequently than this – once every 4–7 days.

Trade names for chemicals

Chemicals used in agriculture have two names: a common chemical name and a trade name. The trade name is the name given to the chemical by the company selling it. If several different companies manufacture the same chemical it may be sold under several different trade names, and this can be confusing.

In this book, each chemical has been referred to by its common chemical name. On pp. 131–5 is a list of these common chemical names and the trade names used for them. By looking up the recommended chemical on this list, you will be able to see which trade names are used for it.

Safe handling and storage of chemicals

The chemicals used to control pests and diseases are often poisonous,

Figure 18   Wear protective clothing when spraying, and keep chemicals locked up and away from children

and careless use can result in sickness and even death. Herbicides (chemicals used to kill weeds) are also very dangerous. All such chemicals must be stored in a safe place, preferably locked up, and out of the reach of children and animals. Do not put chemicals into other packages or bottles – always keep them in their original containers. Empty containers must always be destroyed, *not* used for any other purpose, and not sold in the market, even when washed.

The most serious accidents occur when people drink liquid chemicals by mistake. Deaths have also sometimes occurred because people have drunk pesticides to kill internal parasites. Make sure everyone knows that pesticides are quite different from human medicines, and must never be used to treat illness. When spraying chemicals you must take great care because they are dangerous if incorrectly used. Always follow these rules:

1. Avoid spraying on windy days in case the wind carries the chemical spray to nearby houses or on to other crops.
2. Do not breathe in chemical fumes or dust. Tie a piece of cloth around your face, covering your nose and mouth. Wash the cloth thoroughly afterwards.
3. Wear an overall or old clothes when spraying. Wash them afterwards and throw the washing water away in a pit. The pit should not be near a stream or other water supply: it should be at least 100 metres away from such water sources.
4. Do not eat food or smoke cigarettes when spraying, because there will be some of the chemical on your fingers, which will then get into your mouth.
5. Always wash your hands and face with soap and water after spraying.
6. If any chemicals are spilled on to your hands, wash them immediately. If possible, wear rubber gloves, especially when mixing chemicals.
7. Do not allow anyone to come near while spraying is going on.

Mixing chemicals

In this book, directions for mixing and applying chemicals have not been included. The directions given by the manufacturer on the package should be read very carefully and *followed exactly*. Read the safety rules on this page before mixing chemicals.

Eating fruit and vegetables after spraying

After crops have been sprayed, the fruit or vegetables must not be picked and eaten for some time – usually about two weeks. If the crop is eaten before this time, those who eat it may be poisoned. Vegetables should be washed before being eaten so as to remove any insecticide residue.

Washing spray machines

Chemicals are corrosive and will damage equipment. The residue left in the sprayer could also damage the next crop it is used on. Machines should be washed immediately after use. They should not be washed in a lake or stream as this will pollute the water. Instead, they should be washed over a pit and the water thrown away in the pit afterwards. This pit should be at least 100 metres away from any stream or other water supply.

Storage of dry crops

Damage to stored dry crops such as maize and beans, is mainly caused by mould, insects or rodents. Mould growth occurs when the crop is not dry enough before being stored, so always leave the crop to dry in the sun for several days.

A storage hut should be constructed as follows:

1. It should be raised on legs with the bottom of the store 1 metre clear of the ground and with tin rat-guards at the top of the legs.
2. It should be built with woven sides which permit air ventilation, and so reduce mould damage. The width of the store should not be more than 1 metre and it should face the prevailing wind so as to improve the drying of the crop.
3. It should have a well-thatched or corrugated iron roof to keep the crop dry.

The area around the store should be cleared thoroughly to remove vegetable waste in which weevils and other pests can live. Inside the store, sacks of beans or grain should be raised off the ground to allow air circulation and prevent the crop from 'sweating'.

Figure 19   Food stores need a good roof and should be protected with rat guards

Metal rat guard

Insects may arrive in the store with the crop, or they may already be present in the store. To reduce the number of insects in the store, sweep it out thoroughly before putting the new crop in, and dust the walls with an insecticide such as Pirimephos methyl, Malathion, Permethrin, or any Green Circle compound.

If the crop is to be stored for more than 2–3 months, or if insects are a particular problem, then the crop can be mixed with Pirimephos methyl, Malathion or Permethrin before it is stored. Alternatively, a Blue Cross or Red Triangle compound can be used. Wash the crop thoroughly before cooking to remove the insecticide.

Note: Green Circle, Blue Cross and Red Triangle are the signs put on the packets of insecticides intended for storage pests. The active ingredient is normally Malathion or Pirimephos methyl, but synthetic pyrethroids are sometimes used instead.

The simplest method of mixing the dried crop with insecticide is to put 100 kg of the crop on the floor, sprinkle 125 gm of the insecticide on to it, and mix them up with a shovel. When doing this wear old clothes, and tie a cloth around your face, covering your nose and mouth. Follow the other safety rules given on p. 13

Marketing fruit and vegetables

Fresh fruit and vegetables should reach the market in good condition to fetch high prices. The chief problem is loss of moisture, and to help prevent this you should pick the crop early in the morning, before the dew has dried. Once picked, they should be kept in a cool store and not packed too tightly in their box. Send vegetables and fruit to market as quickly as possible. If this has to be done during the day, the lorry (truck) should be covered, because much of the moisture will be lost if the vegetables or fruit are exposed to hot sunshine. Damage can also be done by rough handling or packing. This bruises the fruit and vegetables, which shortens the length of time they will store and reduces their market value.

Section 2
Growing Fruit and Vegetables

Bananas and plantains
Musa spp

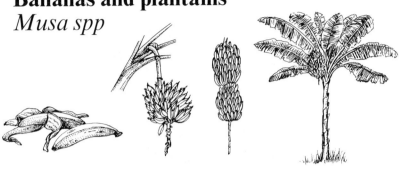

Introduction

The fruit of these plants are either cooked, made into flour, or eaten as fresh fruit. Plantains contain more starch and less sugar than bananas, and are generally cooked before eating, whereas bananas are usually eaten raw. Bananas or plantains can be planted to give some wind protection to a vegetable garden and the leaves are useful for feeding livestock.

World production

Bananas

Brazil	6,424 thousand tonnes per year
Indonesia	2,905
Philippines	2,403
Ecuador	2,391
Thailand	2,082
Mexico	1,929
Burundi	967
Total world production	39,129

Plantains

Total world production	20,584

Climatic range

Bananas require a temperature of 24°C or more for good growth and therefore grow well between sea level and 1,800 metres. They need at least 1,000 mm of rain per year, evenly distributed throughout the year. Moisture should be plentiful at flowering time. They cannot withstand prolonged drought.

Soils

Bananas and plantains should be grown on well-drained, fertile soil; they cannot tolerate waterlogging.

Planting material

The two main varieties are Robusta and Dwarf Cavendish. Robusta is a tall high-yielding variety grown mostly in the Caribbean and the Pacific. The Dwarf Cavendish is a short, hardy banana, grown in Africa and Asia.

There are many different varieties of plantain.

There are several different types of planting material from which bananas and plantains can be grown. Suckers are the best material, but small shoots or parts taken from the rootstock of the parent plant can also be used. Suckers should be taken from the parent plants when they are about 50 cm–2 metres high, and about 15–25 cm wide at the base.

Only take planting material from healthy plants.

Before planting, see control measures for Banana Weevils and Nematodes on pp. 18 and 19.

Spacing

The following spacing should be used when planting out bananas:
For short varieties: 3 metres × 3 metres
For medium varieties: 3 metres × 4 metres
For tall varieties: 4 metres × 4 metres

Bananas are often intercropped with maize, beans or other vegetables.

Planting

Bananas should not be planted on very exposed sites as they will suffer from wind damage.

Plough the land during the dry season and remove all weeds, particularly couch grass. Plant the bananas at the beginning of the rainy season. Dig holes 60 cm deep and 60 cm wide, keeping the topsoil and subsoil separate. Mix the topsoil from each planting hole with a 20-litre drumful of compost or manure, and 100–150 gm double superphosphate (use 1 kg tinful for every 8 plants, or 5–7 large spoonfuls per plant). Half fill the hole with this mixture. Plant the sucker with the base 30 cm deep, and fill up the hole with the rest of the mixture. Make sure that the sucker is firmly planted in the soil. Use the subsoil to make an irrigation ring around the plant.

Fertilizer

Apply 100–125 gm of CAN per plant every year (1 kg tinful for every 8 plants, or 5–7 large spoonfuls per plant). Apply this at the beginning of the rainy season, starting in the second year. Scatter the fertilizer in a circle around the plant.

Weeding

When hoeing, take care not to disturb the roots of the plants. These are found in the top 15 cm of the soil.

Mulching

Mulching is of great benefit as it helps to retain soil moisture and protects the soil from erosion. Mulch also increases the organic matter in the soil as it rots down.

Banana stems cut during thinning can be used as a mulch, but they must be kept several centimetres away from the base of the plant, as a precaution against weevils.

Thinning

The shoots of banana plants should be thinned out, so that the plants produce large fruit. Leave one shoot bearing fruit, one half-grown shoot, and one shoot that is just starting. After harvesting, cut off

Figure 20   After harvesting cut off the old stem close to the ground

the main shoot near the base, leaving the other two shoots more room to develop (see Figure 20).

If the banana leaves and stalks are needed to feed livestock, then the banana plants should not be thinned. The yield will be lower and the fruit smaller as a result.

Staking

Stems bearing heavy bunches of fruit should be supported by sticks or poles (see Figure 21 on page 18).

Harvesting

The first harvest starts 15–18 months after planting. Cut the stem when the bananas are fully developed, and light green and shiny in appearance. Yields of 17–20 tonnes per hectare per year can be obtained.

Transport and storage

When transporting bananas, wrap the bunches in grass or banana leaves to avoid bruising. For temporary storage, keep the bunches in a cool place.

Figure 21   Support heavy bunches with poles

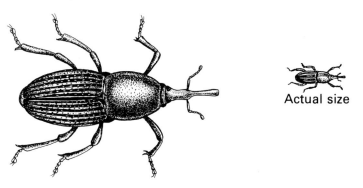

Actual size

Figure 22   Banana Weevil. The weevil is brown when newly emerged, but later turns black

PESTS AND DISEASES	SYMPTOMS	CONTROL
Banana Weevil *Cosmopolites sordidus* (see Figure 22)	Adult weevils are seen on the ground. The white, legless grubs burrow inside the stem. Attacked plants may be stunted, the bunches of fruit small and poor. Badly infested stems are easily blown over.	*If the pest is present:* 1. Cut all old stems off at ground level. Dust the cut rootstock with Dieldrin and cover with soil. 2. Sprinkle Dieldrin, Aldrin, Trichlorphon or Gamma BHC dust around the base of each plant. *As a preventive measure:* Plant clean, healthy suckers. If grubs can be seen on the suckers, or if Banana Weevil is known to be a problem locally, the suckers should be dipped in Dieldrin solution before planting.

PESTS AND DISEASES	SYMPTOMS	CONTROL

Banana Thrips
Hercinothrips bicinctus

These insects are found on the fruit, where they produce silvery or brown patches. If heavily infested, the skins of the fruit may crack.

Spray with Diazinon, Fenitrothion, or Fenthion.

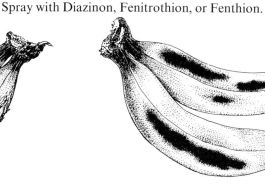

Figure 23   Bananas damaged by Thrips

Nematodes (Eelworms)
Meloidogyne spp

A very serious banana pest. They affect the roots which rot away. The symptoms are brown scars on the roots starting at the tips. The roots become swollen. The plant has few leaves and only produces tiny bananas. These fall off and the plant dies.

If the pest is present:
Abandon the banana plantation. Replant in a new place, using suckers from the old plantation. Dip these in a nematicide before planting.
As preventive measures:
1. Cut back the roots of suckers before planting and dip the roots in a nematicide.
2. Plant on land that has not been used for bananas for at least 5 years previously.

Banana Aphids
Pentalonia nigronervosa

Clusters of brown, shiny, soft-bodied insects found under old leaves near base of stem. This pest spreads Bunchy Top Disease.

Spray stem and around base of stem with Parathion methyl or Demeton-S-methyl.

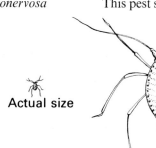 Wingless female
Actual size

 Actual size

Figure 24   Banana Aphids. They are brown in colour and shiny

19

PESTS AND DISEASES	SYMPTOMS	CONTROL
Leaf Spots (Various species of fungi)	Infected patches on the leaves.	*If crop is infected:* Spray with Mancozeb.
Panama Disease *Fusarium spp*	The plants wilt and slowly die.	Burn all infected plants. Obtain new plants from a disease-free plantation. These should be planted in new land, and garden tools should be sterilized to prevent re-infection. Hold the tools in a flame for a few seconds to sterilize them. If possible, plant resistant varieties.
Cigar End Rot *Stachylidium theobromae*	Tips of infected fruit look like cigar ash.	For the next crop, remove the dead flowers from the young banana tips.
Bunchy Top Disease	A viral disease which causes curling and streaking of the leaves. The leaves become smaller, and the fruit are small and misshapen.	Dig up and burn all infected plants as soon as the symptoms are noticed. Spray Aphids to prevent them from spreading the disease.

Beans
Phaseolus vulgaris

common beans

lima beans

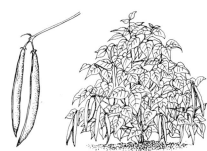

Introduction

This crop is very widely grown, for dried beans or green beans (pods). Both dried and green beans can be sold for canning or export. Dried beans provide much of the protein in the diet of many countries.

Lima beans (*Phaseolus lunatus*) are a useful crop for warmer areas, and are grown in the same way.

Bean plants can help to return some nitrogen to the soil, which improves its fertility.

World production
Dry beans

India	2,240 thousand tonnes per year
Brazil	2,187
Mexico	1,056
Thailand	260
Argentina	232
Burma	185
Rwanda	184
Total world production	14,781

Green beans

Egypt	234 thousand tonnes per year
Chile	81
Syria	46
Thailand	41
India	40
Mexico	29
Total world production	2,527

Climatic range

For green beans a plentiful rainfall is required, unless irrigation is used. For dry beans, the rainfall is not so important; the amount needed depends on the variety. A dry period is essential for ripening. Beans grow best between 1,000 and 2,000 metres above sea level, but they can also be grown satisfactorily outside these limits. They mature faster in warmer areas. Lima beans grow best between 500 metres and 1,500 metres.

Soil

Beans can be grown on a wide range of soils, but best growth is obtained on well-drained loams that are high in organic matter. Bean plants are very sensitive to water-logged soils and standing water will injure the plants in a few hours. Good drainage is therefore essential for bean production.

The soil should be broken up finely before planting.

Seed preparation

Damaged and wrinkled seeds should be thrown out. If using seed kept from the previous crop, it should be dressed with Captan or Thiram before sowning, and with Aldrin or Dieldrin if Bean Fly is a problem (see p. 22).

Beans with dark skins are generally more resistant to attack by fungi in the soil than white-skinned varieties.

Planting

About 50–60 kg of seed will be needed for every hectare of land to be planted.

Dry beans require about 85 days of rain for growth, followed by about 10 days of dry weather for the pods to dry out. They should normally be planted at the beginning of the rainy season, but if this is expected to last much more than 85 days, planting should be delayed, otherwise the beans may rot before they can be harvested. Apply 200 kg/ha DAP along the rows several days before planting (1 kg tinful for every 170 paces along the row, or 1 large spoonful for every 3 paces).

Plant in rows 30 cm apart, with 15 cm between each plant in the row. Plant the seeds 3–5 cm deep. Provide sticks for beans of climbing varieties, and train the beans up the sticks as they grow.

Fertilizer

Apply DAP before planting as already described. When the plants begin to flower apply a top dressing of DAP at 200 kg/ha (1 kg tinful for every 170 paces along the row, or 1 large spoonful for every 3 paces).

Weeding

Keep the field clear of weeds by shallow weeding. Avoid weeding during flowering or when the field is wet, as this can spread disease.

Harvesting

Green beans: Harvesting of the pods begins about nine weeks after sowing and continues for about two months. Pick at regular intervals (about every five days) to maintain quality. Pick only when the weather is dry. After the pods have been picked they should be graded and marketed as soon as possible.

Dry beans: The beans should be harvested when most of the pods have begun to dry out, but before the pods start to shatter. If ripe beans are left exposed to the rain the quality deteriorates rapidly. After threshing and winnowing, sort out rotten, misshapen and damaged beans.

For instructions on storage, see p. 13.

Rotation and field hygiene

Practise crop rotation and avoid having too many overlapping crops of beans, or diseases may become a problem.

After harvesting, clean up all crop residues; plough in, or burn if heavily infected with diseases. Remove any bean plants that have sprung up in the field or around the edge. Make sure that the field is entirely clear of bean plants before the next crop is planted.

Intercropping beans with maize

Provided there is good rainfall, beans intercropped with maize do not reduce the maize yield by very much. This means that a farmer may get more produce from each unit of land by growing maize and beans together.

However, in drier areas, beans do reduce maize yields and do not necessarily make up for this reduction with their own yield. Mixed cropping is therefore not recommended in dry conditions.

Beans with maize: planting

The beans should be planted immediately after the maize. If there is a delay of a week in planting the beans they may yield almost nothing. Plant the maize at the normal spacing and apply the usual amount of fertilizer. Plant the beans in rows, mid-way between the rows of maize, spacing the seed 15 cm apart within the row.

Beans with maize: weeding

Early and thorough weeding is just as important for mixed crops as it is for single ones. Some of the herbicides (weedkillers) that can be used for maize will kill beans, so seek advice before using a herbicide.

Beans with Maize: harvesting

Beans grown with maize are more susceptible to rotting of the pods, especially if damp, cloudy weather occurs when the pods are mature. Under these conditions, the bean plants should be pulled out as soon as the pods turn yellow. They should be kept under cover and then dried in the sun as soon as possible.

PESTS AND DISEASES	SYMPTOMS	CONTROL
Bean Fly *Ophiomyia phaseoli*	The attacked bean seedlings are often yellow, stunted, wilting or dead. Their stems thicken and usually crack just above soil level. Brown–black pupae may be seen.	*If the pest is present:* Spray with Diazinon, Dimethoate, Fenitrothion or Fenthion. *As preventive measures:* 1. If Bean Fly are known to be a problem, dress the seed with Aldrin or Dieldrin before planting. These can be mixed with the normal seed dressing. 2. Commercial growers should routinely spray 7–10 day old seedlings with Diazinon, Dimethoate, Fenitrothion or Fenthion. Repeat the spraying after one week.

Actual size

Figure 25

PESTS AND DISEASES	SYMPTOMS	CONTROL

Bean Aphid
Aphis favae

Small, soft, black insects found in clusters around stems, young shoots and pods, and on the underside of leaves. The leaves are often cupped or distorted, and yellowish.

Spray with Dimethoate, Formothion, Diazinon or Demeton-S-methyl.

Actual size

Figure 26

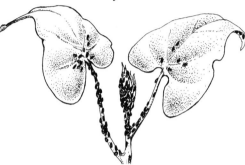

American Bollworm
Heliothis armigera

Young caterpillars feed on flowers, buds and young pods. Older caterpillars burrow into large pods and eat the seeds.

As soon as young caterpillars are seen, spray with one of the following: Fenvalerate, Permethrin, Cypermethrin, Dichlorvos or Trichlorphon, Profenofos, Quinalphos of Methidathion.

Adult

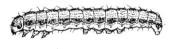
Caterpillar (actual size)

Caterpillar feeding on a bean pod

Figure 27   American Bollworm. The caterpillar is green or brownish with pale bands down its sides

---

Bean Bruchid
Acanthoscelides obtectus

This pest attacks stored beans. The pale yellow or cream-coloured larvae bore into the seeds. Attacked seeds may have dark, circular windows in their skins.

As preventive measures:
1. Sweep out the store thoroughly before use and dust the walls with Pirimephos methyl.
2. If Bean Bruchid tends to be a problem, or if the beans are to be stored for over 3 months, then dust the beans with Pirimephos methyl before storing. Wash well before cooking.

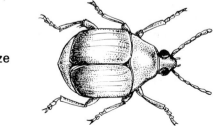
Actual size

Figure 28

PESTS AND DISEASES	SYMPTOMS	CONTROL
Flower Thrips *Taeniothrips sjostedti*	Very small, slender, yellow to black insects which are found concealed in the flowers and jump when disturbed. They suck plant sap from young pods, causing water-soaked spots and curling of the pods. Or the flowers may drop off without forming pods. They are important when growing green beans as they affect pod quality.	Spray with Diazinon, Fenthion, Fenitrothion or Malathion.
Golden Wing Moth *Plusia orichalcea*	The caterpillars, which are green, eat the leaves.	As for American Bollworm.
Rust *Uromyces phaseoli*	Red-brown pimples develop on leaves and sometimes on the pods.	*If crop is infected:* Spray with copper oxychloride or Mancozeb weekly. It is generally uneconomic to spray dry bean crops. *As preventive measures:* 1. If growing green beans for export, a routine spray with copper oxychloride or Mancozeb is recommended. Spray once a week, from emergence. 2. If Rust is a particular problem, plant resistant varieties. Seek local advice on these.
Anthracnose *Colletotrichum lindemuthianum*	Brown marks on pods and stems, brown spots on leaves. The disease appears slowly and gradually spreads. Affected pods are unsaleable.	*If plants are infected:* Spray with Benomyl. *As preventive measures:* 1. Buy certified seed. 2. Plant resistant varieties. 3. Dress seed with Captan or Thiram if not already dressed. 4. In wet weather spray routinely with Benomyl, Mancozeb or a copper fungicide. This will only be worthwhile for a large, valuable crop. 5. Burn crop residues. 6. Practise crop rotation.

PESTS AND DISEASES	SYMPTOMS	CONTROL
Bacterial Blight *Pseudomonas phaseolicola*	Brown, water-soaked marks on pods, sometimes with a 'halo'. This disease is carried by the seeds.	*If plants are infected:* Pull up infected plants and burn. *As preventive measures:* 1. Buy certified seed. 2. Practise crop rotation. Each field should have at least 2 years without beans or peas.
Bean Mosaic Virus	The infected plants are mottled and lose their green colour. In extreme cases the plants are stunted and the leaves misshapen.	*If plants are infected:* Pull up and burn infected plants. Spray Aphids to prevent the disease spreading. *As a preventive measure:* Buy certified seed.
Ashy Stem Blight or **Charcoal Rot** *Macrophomina phaseolina*	Black, sunken cankers with distinct margins, on the stems. Sometimes a mass of tiny bodies, ashy in appearance, can be seen on the surface of the infected stem. In severe cases, the plants wilt and die.	*If plants are infected:* Pull up and burn infected plants. *As preventive measures:* 1. Buy certified seed. 2. Dress seed with Thiram or Captan, if not already dressed.
Angular Leaf Spot *Isariopsis griseola*	Small, brown spots develop on the leaves. They have straight edges formed by the veins of the leaf. Pods may be affected in wet weather.	As for Anthracnose.
Fusarium Root Rot *Fusarium solani* and *F. phaseoli*	Red discoloration of the tap root, which later turns brown. Plants gradually turn yellow.	This disease cannot be controlled by chemical spraying. *As preventive measures:* 1. Practise crop rotation. 2. Dress seed with Captan or Thiram if not already dressed.
Grey Mould *Botrytis cinerea*	The stems, and later the pods, are covered in grey-brown patches which become mouldy in damp conditions. The plants may collapse.	Pull up and burn any seriously infected plants. Remove infected pods from other plants. *As a preventive measure:* Spray with Benomyl, beginning when the plants are forming their first pods.

PESTS AND DISEASES	SYMPTOMS	CONTROL
Sclerotinia *Sclerotinia spp*	Soft, watery marks on young plants, accompanied by white fungus which produces hard, black bodies (sclerotia).	*If plants are infected:* Pull up and burn infected plants. *As preventive measures:* 1. Buy certified seed. 2. Dress seed with Zineb, Maneb or Mancozeb if not already dressed. 3. Practise crop rotation.
Bean Scab *Elsinoe phaseoli*	Brown scabs on pods, leaves and stems.	*As a preventive measure:* Spray with Benomyl or Mancozeb.

Broad beans
Vicia faba

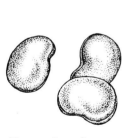

Introduction

The seeds may be used as a fresh vegetable, in which case they are harvested before they become too mature and hard, or they can be dried. Broad beans are also known as horse beans.

World production

Ethiopia	272 thousand tonnes per year
Egypt	236
Morocco	148
Brazil	85
Turkey	52
Mexico	51
Total world production	7,376

Climatic range

Grows only in areas with cool conditions and fairly low humidity. Areas that are 1,000 metres or more above sea level are the most suitable. In warmer areas, lima beans are recommended as an alternative crop (see p. 21).

Soil

Medium to heavy soils are suitable for this crop.

Planting

Sow at the beginning of the rainy season. Sow the seed directly into the vegetable beds at a depth of 5 cm. They should be in rows 50 cm apart, with 15 cm between each plant in the row. Sow a small amount of seed at monthly intervals to give a regular supply of fresh beans.

Manure and fertilizer

If possible, dig manure into the soil before planting. Several days before planting, apply 200 ka/ha of DAP (1 kg tinful for every 100 paces along the row, or 1 large spoonful for every 2 paces).

Cultivation

When the plants are about a metre tall, pinch out the growing point, as shown in Figure 29. This will prevent the plant from growing any taller and encourage large, early pods.

Harvesting

The beans are ready to harvest 12–16 weeks after planting. For fresh beans, pick before the pods become hard. For dried beans, allow the pods to mature completely.

Figure 29   Pinching out the growing point of a broad bean plant

PESTS AND DISEASES	SYMPTOMS	CONTROL
Aphids (Various species) (see Figure 26 on p. 23)	Aphids attack the young shoots and leaves and cause distortion and sometimes yellowing of the leaves.	Spray with one of the following: Dimethoate, Formothion or Diazinon.
Chocolate Leaf Spot *Botrytis cinera*	Brown spots on leaves.	This disease cannot be effectively controlled by chemical spraying. *As a preventive measure:* Plant the crop earlier next time, so that it is well established before the heavy rains begin.

Brassicas
Brassica spp

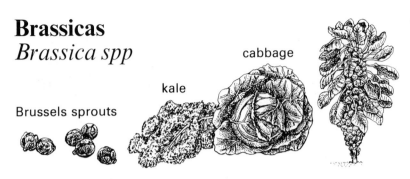

Brussels sprouts

kale

cabbage

Introduction

This family includes cabbage, kale, Brussels sprouts and cauliflower. As they are grown in much the same way, and suffer from the same diseases, they are covered under this one heading.

These vegetables are usually eaten boiled, although cabbage can be eaten raw as a salad. The cauliflower is grown for its white head, which is made up of clusters of unopened flowers. Brussels sprouts are cultivated for the enlarged buds that grow from the main stem.

Word production

Cabbage

Korean Republic	1,034 thousand tonnes per year
Turkey	566
India	450
Colombia	449
Egypt	360
Korea DPR	320
Total world production	33,560

Climatic range

Brassicas grow well under cool, moist conditions. They do not like hot temperatures, and grow best at altitudes of 700 metres or more.

Kale and cabbage are more tolerant of dry conditions than cauliflower or Brussels sprouts. Cauliflower and Brussels sprouts will only grow successfully in colder, highland areas.

Varieties

Cabbage

Sugar Loaf	An early-maturing variety, taking $2\frac{1}{2}$ months from sowing to harvest.
O.S. Cross Prize Drumhead Copenhagen Market Glory of Enkhuizen Early Jersey Wakefield	All these are mid-season varieties, taking about $3\frac{1}{2}$ months from sowing to harvest.

Kale

Thousand-headed	A very popular variety with succulent leaves.
Marrow-stem	A tall, vigorous variety, with dark green, finely curled leaves of good flavour. Less prone to bird damage than other varieties.

Brussels sprout

Covent Garden	A thickly-set variety with compact sprouts of delicate flavour and attractive appearance.
Jade Cross	An F_1 hybrid variety that is very early and uniform. The sprouts are medium-sized and compact.

Cauliflower

Early Giant	A variety with a large, compact, well-protected head.
Extra Early Six-Week	Early, small, round-headed variety.
Snow Giant	A compact variety.
Patna Early	A small variety suitable for warmer areas.

Planting

About 500 gm of seed is required for every hectare of land to be planted. For kale, 280 gm per hectare is needed. Brassicas should be started in the nursery. Make raised beds about a metre wide and of convenient length. Apply manure and double superphosphate and work well into the soil; use about 200 kg/ha of double superphosphate (1 large spoonful per square metre).

Make drills 20 cm apart, sow the seeds thinly, and cover them lightly with soil. Later, thin out to 2–3 cm apart, to give strong seedlings. Water once or twice a day, and in hot areas shade the beds. Too much shade, too much water or overcrowding, will all encourage attack by mildew. Keep a look-out for aphids and caterpillars.

Transplanting

Transplant seedlings at about one month of age, when they have four true leaves (see Figure 12 on p. 7). They should be about 12 cm tall. Transplant when the soil is moist, and choose a cool time of day – the evening or early morning.

Firm the plants down well and water in. With cauliflowers, it is essential that no check in growth occurs, or the plant may refuse to head properly, so make sure they always have plenty of water.

Spacing

Space large varieties of cabbage, and all types of kale and cauliflower at 60 cm × 60 cm. Small varieties of cabbage can be spaced at 45 cm × 45 cm. Brussels sprouts should be spaced at about 75 cm × 75 cm.

Manure and fertilizer

If available, dig manure into the ground before planting. Add one large spoonful of double superphosphate to each planting hole (this is equivalent to 550 kg/ha if the plants are spaced at 60 cm × 60 cm).

When the seedlings are 20 cm high, top-dress with CAN, applying one large spoonful per plant (550 kg/ha). Three weeks later, top-dress again with the same amount of CAN.

Weeding

Keep the field free of weeds until the crop's leaves cover the ground.

Harvesting

Brussels sprouts and cabbage: Harvest when the heads become firm.

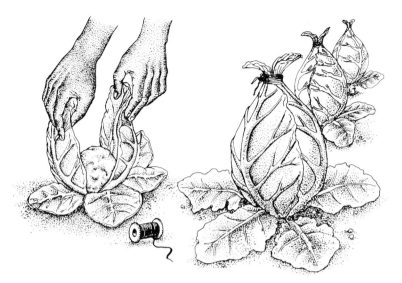

Figure 30   Tying up the leaves of a cauliflower to protect the curd

Figure 31   Harvesting kale

Pick sprouts from the bottom of the plant first. Yields of about 20 tonnes per hectare are generally obtained.

Cauliflower: As soon as the white curd can be seen, fold the leaves upwards and tie the tips together with string as shown in Figure 30. This will prevent discolouration of the curd.

Cut the head when it is mature and firm. Cut it so that there are a few leaves around the curd, to prevent it from damage when being handled. Market in one or two days, while the heads are still fresh.

Kale: Pick the lower leaves when they are ready (see Figure 31). Pick regularly and use them or sell them as soon as possible, before they wither.

Field hygiene

Remove the stumps of the plants from the field to prevent the carry-over of disease. If seriously infected by disease they should be burned.

PESTS AND DISEASES	SYMPTOMS	CONTROL
Diamond Back Moth *Plutella maculipennis*	The pale green caterpillars eat the underside of the leaf, making holes right through it.	Spray with Fenvalerate, Cypermethrin, Carbaryl or Permethrin.

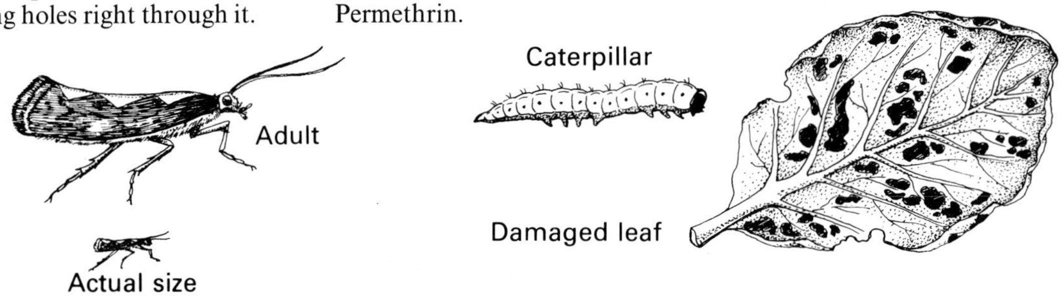

Adult

Caterpillar

Damaged leaf

Figure 32

Actual size

| **Cabbage Sawfly**
Athalia spp | The caterpillars eat the leaves, often leaving only the midrib. (Affects all Brassicas, not just cabbage.) | Spray with Fenvalerate, Permethrin, Cypermethrin, Carbaryl, or Malathion. |

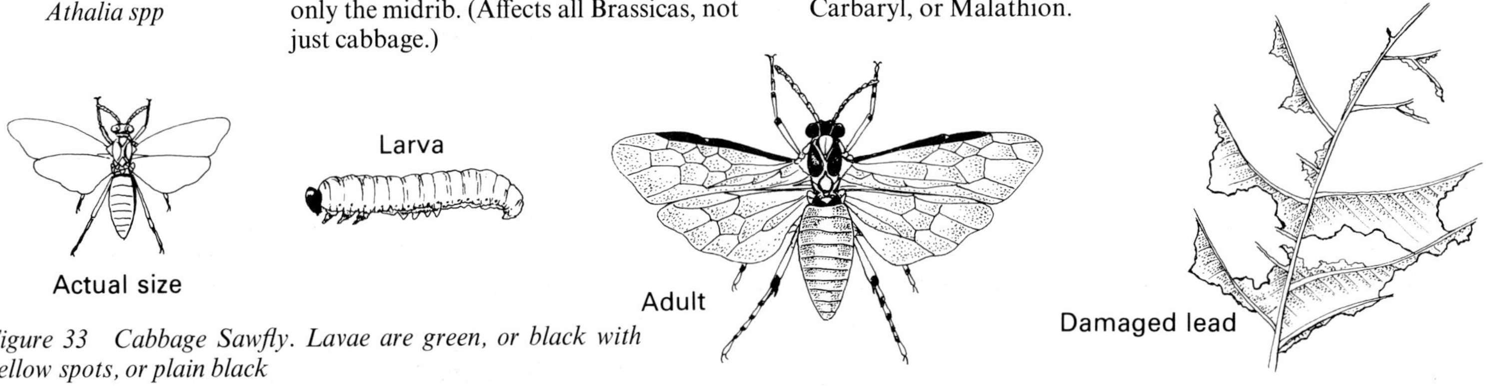

Larva

Actual size

Adult

Damaged lead

Figure 33   Cabbage Sawfly. Lavae are green, or black with yellow spots, or plain black

PESTS AND DISEASES	SYMPTOMS	CONTROL
Cabbage Aphid *Brevicoryne brassicae*	Small, pale green insects, covered with a light dust of mealy powder. (Affects all Brassicas, not just cabbage.)	Spray with Diazinon, Formothion or Dimethoate.
Cut-worms *Agrotis spp*	Grey to black caterpillars which feed at night. They bite out the stem at ground level, causing the plant to fall over. They can be found hidden in the soil near the cut seedlings.	1. Dust around the plants with an insecticide, such as Aldrin or Dieldrin. 2. Make up a bait, with 10 kg bran, 500 gm sugar and 10 litres water. Add either 100 gm Trichlorphon (95% sp) or 100 gm Aldrin (40%wp) to the dry bran and mix in well. Add the sugar and water. This produces a crumbly mass. Sprinkle it on the ground around the plants.

Caterpillar

Adult

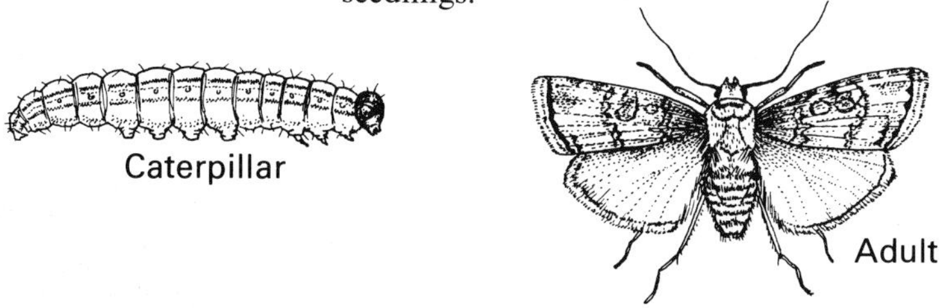

Figure 34   Cut-worm (actual size). Pale brown with faint dark lines. The moth is brown with whitish hindwings

Flea Beetles *Phyllotreta spp*	The adults feed on the leaves of young plants, and sometimes kill the seedlings. The larvae feed on the roots of the plants, but do not do much damage.	If the plants are severely infested spray with Permethrin, Diazinon or Carbaryl.

Two different types of Flea Beetle

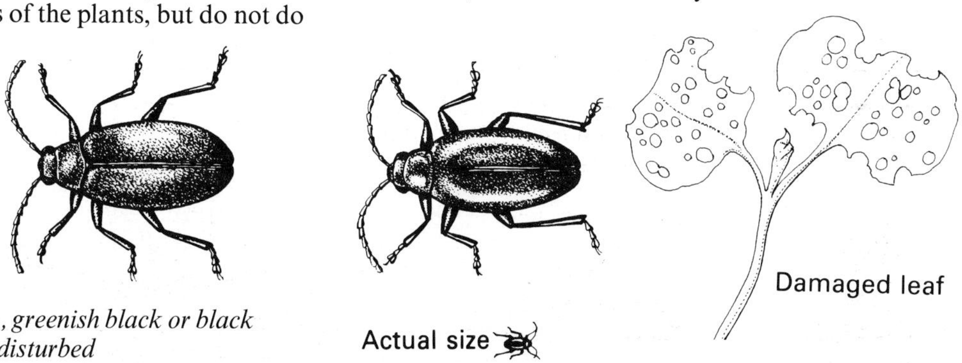

Damaged leaf

Actual size

Figure 35   Flea Beetles. The beetles are black, greenish black or black with yellow stripes. They jump like fleas when disturbed

Slugs	Large holes in leaves. Slugs hidden under leaves.	Scatter slug pellets around the plants. Or you can bury a tin in the ground, with its rim level with the soil surface. Fill the tin with beer. The slugs are attracted by the smell of the beer and drown.

PESTS AND DISEASES	SYMPTOMS	CONTROL
Black Rot *Xanthomonas campestris*	Stems and roots go rotten. The rot gives off a characteristic bad smell. A serious disease that can cause total crop loss in fields with overhead irrigation. Infection can occur at all stages of growth.	This disease cannot be controlled by chemical spraying. Burn all infected plants. *As preventive measures:* 1. Buy certified seed. 2. Practise crop rotation.
Dry Rot Canker or **Black Leg** *Phoma lingam*	Caused by a fungus carried in the seeds. Brown to black spots on the seedlings while in the nursery. Old spots are ash-grey and often have small pimples. A dark-coloured stem canker extends below the soil level, killing the roots. Mature plants wilt suddenly and die.	This disease cannot be controlled by chemical spraying. Burn all infected plants. *As preventive measures:* 1. Buy certified seed. 2. Practise crop rotation.
Ring Spot *Mycosphaerella brassicicola*	Brown spots on the leaves, often with a green border. Seed-borne disease that is also spread by wind or compost made from infected material.	*If plants are infected:* Burn all infected plants. Spray with Benomyl to prevent the disease spreading. *As preventive measures:* 1. Buy certified seed. 2. Only transplant healthy seedlings from the nursery. 3. Practise crop rotation.
Dark Leaf Spot *Alternaria spp*	Small spots on leaves of seedlings. If infection is severe there may be 'damping off' of seedlings (i.e. the stem goes soft near the ground and the seedling collapses). On mature leaves, dark brown circular spots, 1–2 cm across, appear. The leaves eventually dry up. This disease is seed-borne, but it is also spread by wind, water and garden tools.	This disease cannot be controlled by chemical spraying. *As preventive measures:* 1. Buy certified seed. 2. Only transplant healthy seedlings from the nursery. 3. Practise crop rotation.
Downy Mildew *Peronospora parasitica*	A fluffy fungal growth on the undersides of leaves. Later, brown to black spots appear on upper surface. The disease is mainly spread by wind and rain. Severe at high altitudes where conditions are cool and wet, or where the nursery is over-shaded or over-watered.	*If plants are infected:* Spray with a fungicide such as Benomyl. *As preventive measures:* 1. Practise field hygiene in the nursery. 2. Practise crop rotation.

PESTS AND DISEASES	SYMPTOMS	CONTROL
White Rust *Cystopus candidus*	White, blister-like pimples on the leaves, stems and flowers, and sometimes on the seed pods.	This disease cannot be controlled by chemical spraying. Burn all infected plants. *As a preventive measure:* Practise crop rotation.

Brinjals
Solanum melongena

Introduction

The brinjal is grown for its fruit. It is a useful crop for hotter areas and has a high export potential.

Brinjals are also known as aubergines or egg-plants.

World production

Turkey	550	thousand tonnes per year
Egypt	290	
Syria	148	
Iraq	116	
Philippines	95	
Jordan	64	
Total world production	4,273	

Climatic range

Brinjals prefer warm to hot areas and grow well between sea level and heights of 1,500 metres. They may require irrigation in some areas.

Varieties

Florida High Bush is an ideal variety to grow for export. Other good varieties are Black Beauty, which has a longer storage life than average, and Early Long Purple, which is high yielding.

Sowing

Brinjals germinate poorly when sown directly into the field, so they should be started in a nursery. Prepare the beds and sow the seeds thinly in rows 30 cm apart. Much with fine grass. Later thin out the seedlings to 5 cm apart.

Transplanting

Transplant when the seedlings are strong and have four true leaves (see Figure 12 on p. 7). Space plants in rows 90 cm apart, with 60 cm between each plant in the row.

Manure and fertilizer

Mix 1–2 handfuls of manure and one large spoonful of double superphosphate with the soil in each planting hole. (This is the equivalent of 370 kg/ha of double superphosphate.)

When the plants are 25 cm high, top dress with CAN, applying one large spoonful for every plant (370 kg/ha). Three weeks later apply two large spoonfuls of CAN per plant (740 kg/ha).

Mulching

Mulching is recommended, if there is mulch available. This keeps the fruit clean of rain and irrigation splashes.

Weeding

Keep the field free of weeds until the crop's leaves cover the ground.

Harvesting

Brinjals bear ripe fruit after five months and the harvest continues for a further two to four months.

Harvest when the fruits are dark glossy purple, and before they turn yellow. Cut the stalk, leaving a small piece of stalk attached to the fruit.

Handle brinjals carefully to avoid bruising.

PESTS AND DISEASES	SYMPTOMS	CONTROL
Epilachna Beetle or **Vegetarian Ladybird** *Epilachna spp*	The beetles eat the leaves in patches, leaving a skeleton of small veins.	Spray with Fenthion, Fenitrothion, Malathion, Carbaryl or Trichlorphon.
Striped Blister Beetle *Epicauta albovittata*	The black-and-white striped beetles eat irregular holes in the leaves.	Spray with Parathion methyl.

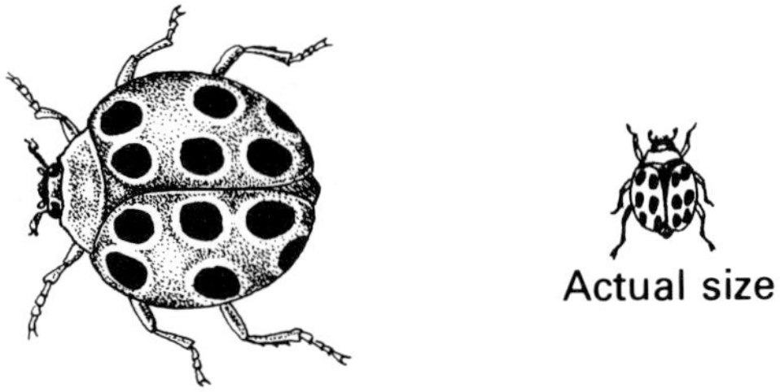

Figure 36   Epilachna Beetle. Red with black spots. This beetle eats the leaves of crops. Similar beetles are beneficial as they eat aphids, so check for leaf damage before using insecticide

Actual size

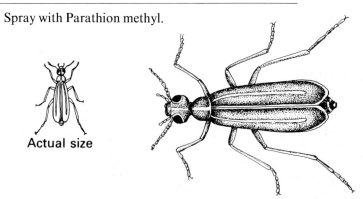

Figure 37   Striped Blister Beetle. Black with white stripes

Actual size

PESTS AND DISEASES	SYMPTOMS	CONTROL
Gall Midge *Asphondylia spp*	The larvae feed inside the fruit and shoots, causing galls.	Spray with Malathion, or Fenthion. Commercial growers should spray routinely, starting at flowering stage. Other farmers will probably not find this economic, and should spray only if the pest is present.
Potato Tuber Moth **Phthorimea** **operculella** (see Figure 70 on p. 80)	The caterpillars eat the leaves.	Spray with Dimethoate.
Bacterial Wilt *Pseudomonas* *solanacearum*	The plants wilt and die. This disease is likely to be serious if brinjals are planted on land that has previously had on it potatoes, tomatoes, peppers, tobacco or other crops of this family, infected with this disease.	This disease cannot be controlled by chemical spraying. Pull up and burn any infected plants. *As preventive measures:* 1. Buy certified seed. 2. Practise crop rotation. Each field should have at least 4 years without crops of this family.
Leaf Spot *Alternaria* *solani*	Circular brown spots appear on the leaves.	Once this disease is established spraying is not worthwhile. *As a preventive measure:* Spray with Mancozeb every week, starting from emergence. This will probably only be worthwhile for commercial growers.

Carrots
Daucus carota

Introduction

This crop is grown for its edible root, which is eaten cooked, or raw in salads. The roots can be stored for several months and can also be dried or canned.

Long radish, also known as Chinese radish or white carrot, is cultivated in the same way.

World production

Colombia	153 thousand tonnes per year
Egypt	111
Argentina	90
Mexico	72
Chile	52
Turkey	50
Total world production	10,325

Varieties

Chantenay is recommended for the fresh market, while Nantes is a good variety for both the fresh market and canning. For livestock feeding, the variety Oxhart is useful.

Recommended varieties of long radish are Lobak and Hong Kong Summer.

Climatic range

A cool to warm climate is required. Carrots are very sensitive to high soil temperatures, which will cause pale, short roots to be produced.

Long radish tolerates a wide range of climatic conditions.

Fertilizer

Do not use manure as it causes the roots to fork.

Before planting, apply 200 kg/ha of double superphosphate (1 kg tinful for every 170 paces along the row, or 1 large spoonful for every 3 paces). If double superphosphate was applied for the previous crop then this will not be necessary.

When the plants are 10 cm high, apply CAN at 200 kg/ha (1 kg tinful for every 170 paces along the row, or 1 large spoonful for every 3 paces).

Sowing and thinning

About 5·5 kg of seed per hectare will be needed.

Sow directly into the field. Sow thinly along drills, which should be 30 cm apart.

About 4 weeks after sowing thin out; for canning leave 4 cm between plants and for the fresh market leave 6–7 cm between plants. For drying leave 5 cm between plants, and thin a second time when the carrots are fairly large, to leave 10–15 cm between plants. The carrots pulled up at the second thinning can be eaten.

Harvesting

The crop matures 3–4 months after sowing. For canning, tender young roots are required, so harvest early. Carrots intended for drying should be allowed to grow beyond fresh-market size to increase the yield.

Yields

Yields depend on the stage at which the roots are lifted. For canning, yields of about 20 tonnes/ha are possible. Fresh carrots can yield about 25 tonnes/ha, while for drying yields of up to 50 tonnes/ha can be obtained.

PESTS AND DISEASES	SYMPTOMS	CONTROL
Aphids *Myzus spp*	Green aphids found at the base of the leaves.	Spray with Diazinon, Formothion or Dimethoate.
Flea Beetles *Haltica pyritosa* (see Figure 35 on p. 31)	The beetles eat the leaves, and may kill young plants.	If damaged leaves are seen, spray with Malathion, Diazinon, Permethrin or Carbaryl.

Cocoyams
Colocasia antiquorum and *Xanthosoma sagittifolium*

Introduction

There are two species of cocoyams, Colocasia and Xanthosoma, but they are very similar. The corms of cocoyams are boiled whole for several hours and then peeled and eaten; they may also be cut up and cooked with other food.

Cocoyams have a number of different names, including taro, eddo, tannia and dasheen.

Climatic range

Cocoyams require an average rainfall of over 1,200 mm per year, and they are usually grown at altitudes of 900 metres – 1,800 metres.

Colocasia can withstand continued waterlogging, but Xanthosoma cannot tolerate such conditions.

Soil

A deep, fertile soil is required for cocoyams. They grow well along stream banks, and Colocasia can be grown in swamps.

Planting

Side corms should be obtained from disease-free plants for the initial planting. Space the corms 50 cm to 1 metre apart. Alternatively, 3 corms can be planted close together and a space of 1 metre left between the planting stations. Manure or compost can be added to the planting hole before planting. Little weeding is necessary because the plants grow very strongly.

Harvesting

Corms are normally dug up as required. The first corms are ready for harvesting 6–9 months after planting. Harvesting and replanting are done at the same time. The whole plant is dug up and mature corms are cut off, leaving about 5 cm of corm still on the plant. The leaves are then trimmed back and the plant is replanted in the same hole. Manure or compost can be added to the hole and mixed into the soil before replanting.

Yields are normally about 10–15 tonnes per hectare.

PESTS AND DISEASES	SYMPTOMS	CONTROL
Nematodes (Eelworms) *Meloidogyne spp*	Lumps and knots develop in the corms.	There is no worthwhile chemical control for this pest. *As a preventive measure:* Practise crop rotation.
Cocoyam Root Rot *Pythium spp*	The corms become rotten, and the yield is seriously reduced.	There is no chemical control for this disease once it is established. *As preventive measures:* 1. Dust planting material with Captan. 2. Practise crop rotation.

PESTS AND DISEASES	SYMPTOMS	CONTROL
Leaf Blight *Phytophthora colocasiae*	Dark, wet spots appear on the leaves and the plant eventually dies. This disease is prevalent in wet weather.	Spray with Zineb.

Cow peas and long beans
Vigna spp

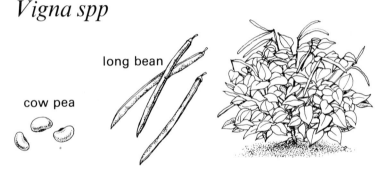

long bean

cow pea

Introduction

Cow peas (*Vigna unguiculata*) are grown for their leaves, pods and seeds. The dried seeds can be cooked in soups or stews, or boiled with maize. The leaves and pods may be fried or boiled, and the leaves are sometimes dried and stored.

Cow peas are an important crop in marginal rainfall areas; they may also be grown for animal feed. Cow peas are sometimes known as black eye peas beans, marble peas, tonkin beans or Chinese peas.

Long beans (*Vigna sesquipedalis*) are very closely related to cow peas. They are primarily an Asian crop and are mainly grown for the green pods, although they can also be grown for the dry seed. This crop is cultivated in much the same way as cow peas.

Long beans are also known as snake beans, yard-long beans, asparagus beans or pea beans.

Climate

Cow peas require warm climates and only give good yields below 1,500 metres. They will survive in very dry conditions and so do well in marginal areas, where they yield better than common beans (*Phaseolus vulgaris*). In wet areas there is considerable insect damage which reduces the seed yields; in such areas they are often grown principally for their leaves.

Long beans require rather more rainfall than cow peas, and 1,600 mm per year gives the highest yields. They grow well at altitudes up to 1,800 metres.

Soil

A well-drained soil is essential for both cow peas and long beans. Apart from this they tolerate a wide variety of soils.

Planting

Cow peas need a dry period when maturing so they should be planted two months before the rainy season ends, or at the beginning of the rainy season if this lasts for less than two months.

Long beans should be planted at the beginning of the rainy season.

Spacing

The spacing used for cow peas can vary considerably. A close spacing has the advantage of giving ground cover very rapidly; this supresses weed growth so that less weeding is needed. However, a large amount of seed must be planted to give a close spacing, so a wider spacing is often used. A spacing of about 45 cm × 45 cm should give reasonable results. The trailing varieties are spaced out more widely than the bush varieties. Both cow peas and long beans are often intercropped (see below), and the spacing is then adapted accordingly.

Cow peas may also be broadcast; this is often done in high rainfall areas, and where they are grown for animal feed.

When long beans are grown alone, they are usually spaced at 60–90 cm × 50 cm. They can also be grown in double rows, as shown Figure 38.

Varieties

There are many different varieties of cow peas and the seeds may be white, cream, purple, brown or black. Some varieties are trailing or spreading types, while others are bush types.

As with cow peas, there are many different local varieties of long beans. Improved varieties, such as Yard Long and Hong Kong Long White are available.

Figure 38   Long beans planted in a double row to make staking easier

Intercropping

Cow peas are often intercropped with yams, maize, millet or sorghum.

Long beans are generally intercropped with sweet potatoes, yams or cocoyams (*Xanthosoma* type).

Fertilizer

Before planting, apply double superphosphate at a rate of 100 kg/ha (1 large spoonful for every 4–5 paces along the row.)

A few weeks later, top dress with 100 kg/ha of ammonium sulphate and 100 kg/ha of muriate of potash.

Alternatively, apply a compound fertilizer with a low nitrogen content before planting.

Cultivation

Weed thoroughly for the first six weeks after planting. Weeding should be done by hand.

Long beans will require stakes for support. Wooden poles, 2 metres long should be provided (see Figure 38).

Harvesting

The leaves of cow peas should be harvested when they are young and tender. The leaves can be harvested at least three times, without affecting the seed yield. They should be picked at weekly intervals. After spraying with chemicals, the leaves must not be picked for two weeks.

The seeds of cow peas mature about 3–5 months after planting. The pods can be picked at intervals, as they ripen, or all at once when most of them are ripe. They should be dried in the sun before being threshed and stored. The average yield is 350–700 kg per hectare, but with good management much higher yields than this can be obtained.

Long bean pods are ready to be harvested 2–3 months after planting. The harvest continues for about 6 weeks and the pods should be picked 2–3 times each week.

If cow peas are grown for the green pods they should be harvested in the same way.

PESTS AND DISEASES	SYMPTOMS	CONTROL
American Bollworm *Heliothis armigera* (see Figure 27 on p. 23).	The caterpillar feeds on the flower buds, flowers and pods.	As soon as young caterpillars are seen spray with Fenvalerate, Permethrin, Cypermethrin, Dichlorvos, Trichlorphon, Quinalphos or Methidathion.
Maruca Moth or **Pod Borer** *Maruca testulalis*	The caterpillars feed on the flower buds and beans.	If the pest is prevalent, spray routinely with Fenvalerate, Permethrin, Cypermethrin, Quinalphos or Methidathion, starting when the plants begin to flower.
Pollen and Blister Beetles *Coryna* and *Mylabris spp*	These beetles eat the flowers, which reduces the yield.	Spray with Fenthion, Fenitrothion, or Trichlorphon.

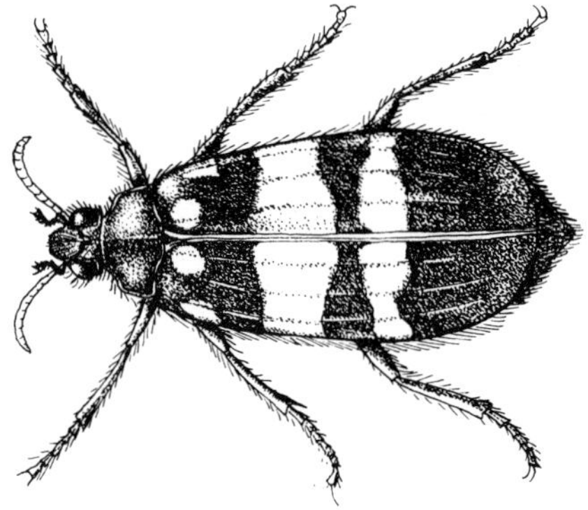

Figure 39   Pollen Beetle (twice life size). Black and yellow, or black and red. If handled produces a yellow liquid which burns the skin

Bean Fly *Ophiomyia phaseoli* (see Figure 25 on p. 22)	This pest attacks the seedlings which are often yellow, stunted, wilting or dead. Their stems thicken and crack just above soil level. Brown-black pupae may be seen.	*If the pest is present:* Spray with Diazinon, Dimethoate, Fenthion or Fenitrothion. *As preventive measures:* 1. If Bean Fly are a problem, dress the seed with Aldrin or Dieldrin before planting. These can be mixed with the normal seed dressing. 2. Commercial growers should routinely spray 7–10 day old seedlings with Diazinon, Dimethoate, Fenthion or Fenitrothion. Repeat the spray after one week.

PESTS AND DISEASES	SYMPTOMS	CONTROL
Spiny Brown Bugs *Acanthomia spp*	The bugs suck the flower buds and pods. Most damage is done by them piercing the pods and seeds.	Spray with Diazinon, Trichlorphon, Malathion, Fenthion or Fenitrothion.
Nematodes (Eelworms) *Meloidogyne spp*	This pest attacks the roots of the plant, which may wilt and die if severely infested. The roots are distorted and lumpy.	There is no worthwhile chemical control for this pest. *As preventive measures:* Practise crop rotation. Land which is severely infested should be planted with grass or cereals for a few years.
Cowpea Weevil *Callasobruchus spp* and other Weevils	The weevils lay their eggs on the pods while the plants are still in the field, or on the stored beans. The larvae burrow into the beans leaving round holes.	If the pest is found on the plants, spray with Malathion. If the pest is prevalent, dust the beans with insecticide before storage. *As a preventive measure:* Harvest the pods as soon as they are ripe. This gives the Weevils less chance to lay their eggs on them.
Thrips	These tiny pests (see Figure 47 on p. 50) feed on the flowers which may fall off or be distorted.	If Thrips are prevalent, spray routinely with Gamma BHC, starting when the flower buds first appear, and spraying once a week for a period of 6 weeks.
Leaf and Pod Spot *Ascochyta phaseolorum* or *Dactyliophora tarii*	Brown spots develop on the pods and leaves.	There is no worthwhile chemical control for this disease. *As a preventive measure:* Plant resistant varieties.
Scab *Synchytrium dolichi*	Raised, warty brown scabs on the stems and pods.	As for Leaf and Pod Spot.
Fusarium Wilt *Fusarium spp*	The plants wilt even though the soil is moist. The roots turn red or brown and the plants eventually die.	This disease cannot be controlled by chemical spraying. *As preventive measures:* 1. Practise crop rotation. 2. Dress seed with Captan or Thiram if not already dressed.

PESTS AND DISEASES	SYMPTOMS	CONTROL
Powdery Mildew *Erysiphe polygoni*	White mildewy growth found underneath the leaves.	Spray with Dinocap, Benomyl or Triadimefon.
Anthracnose *Colletotrichum* *spp*	Brown marks on pods and stems, and brown spots on leaves. If affected the green pods cannot be sold as a fresh vegetable, so this is a serious disease in long beans. The disease appears slowly and gradually spreads.	*If plants are infected:* Spray with Benomyl. *As preventive measures:* 1. Buy certified seed. 2. Plant resistant varieties if these are available. 3. Dress seed with Captan or Thiram if not already dressed. 4. In wet weather, spray routinely with Benomyl, Mancozeb, or a copper fungicide. This will only be worthwhile for a large, valuable crop. 5. Burn crop residues. 6. Practise crop roration.
Mosaic Virus	The leaves become mottled, yellow or crinkled. The yield is reduced.	Dig up and burn all infected plants as soon as they are seen. Spray insects to prevent them from spreading the disease.

Cucurbits
Cucurbitaceae

pumpkin

squash

cucumber

marrow

sweet melon

Introduction

This family of plants includes the cucumber (*Cucumis sativa*) sweet melons (*Cucumis melo*), watermelon (*Citrullus lanatus*) and pumpkins and squashes, (including marrows and zucchini courgettes), all of which are species of *Cucurbita*. Gherkins, gourds and calabashes also belong to this family, as does the loofah (*Luffa cylindrica*). The cucurbits are useful vegetables to grow in warmer areas. They all have similar cultural requirements and generally suffer from the same diseases, so they are dealt with together.

World production

Watermelons			*Cucumbers and gherkins*			*Sweet Melons*	
China	4,090 thousand tonnes per year		China	2,570 thousand tonnes per year		China	1,420 thousand tonnes per year
Turkey	4,000		Turkey	450		Egypt	266
Egypt	1,344		Egypt	231		Syria	212
Iran	930		Syria	196		Chile	130
Iraq	648		Indonesia	154		Total world production	6,372
Brazil	492		Mexico	100			
Mexico	260		Total world production	10,147			
Total world production	24,096						

Climatic range

Cucurbits like warm to hot conditions but need a fairly high rainfall. They grow well in low-rainfall areas under irrigation. Most cucurbits do well up to 1,000 metres, while cucumbers can be grown at higher altitudes, up to 1,500 metres.

Varieties

Cucumber

Palomar	A variety with dark green fruit, slightly pointed at the end. This variety is supposed to be resistant to Downy Mildew.
London Long Green	A white-fleshed variety of excellent taste.
Cool and Crisp	This variety has uniform dark green fruit which taper towards the end.
Early Fortune	A pure white variety with firm, crisp flesh and a small seed cavity.
Colorado	This variety has a long, dark green fruit.

Sweet melon

Recommended varieties are Hale's Best and Honeydew.

Watermelon

Charleston Grey	This variety has large fruit, cylindrical in shape. The rind is tough. The flesh is bright red, crisp and sweet.
Chilean Black	This variety has oval fruit with high-quality, red flesh.
Congo	The fruit of this variety have a rough rind, while the flesh is firm and fine-grained.
Fairfax	The fruit has a thin rind with stripes of light and dark green. This variety is resistant to Anthracnose and Fusarium Wilt.
Sugar-baby	This variety has round fruits, about 20 cm in diameter, which are an ideal size for marketing. The rind is hard and dark in colour, and the medium-red flesh is very sweet.

Pumpkin

Queensland Blue, Green Hubbard and Little Gem are all recommended.

Squash

Long White Bush and Long Green Bush are good varieties of vegetable marrow. The variety Zucchini is used when growing for courgettes.

Intercropping

Cucurbits, especially pumpkins and gourds, are often intercropped with maize. This reduces fruit-loss caused by flower-sucking insects such as Melon Flies.

Manure and fertilizer

At each planting station, dig in a 20-litre drumful of manure and one large spoonful of double superphosphate, before planting. When the plants start to spread, top dress with one large spoonful of CAN per plant. When the plants start to flower apply CAN again, at a rate of one large spoonful per plant.

Planting

Seeds can be sown directly into the field, or into small plastic pots, or in the nursery. Sow the seeds 3–5 cm deep. if they are not sown direct, the seedlings should be transplanted when they are 16 cm high.

Cucurbits are usually grown on mounds or ridges, about 10–12 cm high, as shown in Figure 76 on p. 93. This gives the plants better soil to grow on. Ridges can be made mechanically, using a ridger. Mounds are used when cultivation is done by hand. The fertilizer and manure is added to the soil and mixed in well before the ridges or mounds are made.

Spacing

The spacing varies depending on which type of cucurbit is being grown. The spacing for different types is shown in the table below.

For most cucurbits more than one plant is allowed to grow at each planting station, as there is plenty of room for them to spread out, down the sides of the ridge or mound. The number of plants required at each planting station is shown in the last column of the table below.

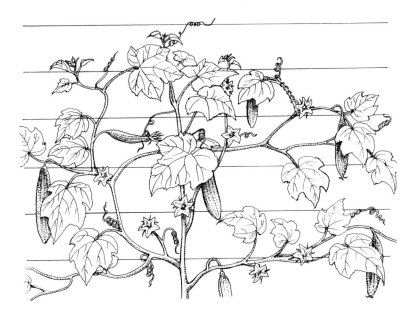

Figure 40   A cucumber vine growing on wires

Crop	Planting distance	Number of seeds in each hole	Number of plants at each station after thinning
Cucumber	90 cm × 60 cm	3–4	2
Squash (including Marrow and Courgettes) (Zucchini)	90 cm × 60 cm	2	1
Sweet Melon	120 cm × 120 cm	4	3
Watermelon	3 metres × 2 metres	6–9	2–3
Pumpkin (trailing)	4 metres × 4 metres	5–6	3
Pumpkin (bush)	120 cm × 120 cm	5–6	3

44

Watering

Regular watering is required, but excessive watering encourages fungal diseases, such as Downy Mildew and Stem Rot. The use of a pot buried in the ground for watering, as shown on Figure 1 on p. 2, is recommended if Stem Rot is prevalent. This method of watering prevents the stem of the plant from getting wet.

Protecting the fruit

If the fruit rests on the ground it may be damaged by damp, or by pests in the soil. This can be avoided by raising the fruit off the ground. Melons, pumpkins and large squashes can be placed on dry grass or straw. With cucumbers, it is best to train the plants along a framework of sticks or wires, so that the fruit grow suspended in the air, clear of the ground, (see Figure 40).

Harvesting

Cucumbers are ready to harvest 6–8 weeks after sowing and yield 30–50 tonnes per hectare. Pick regularly by cutting the stalk with a sharp knife. Do not allow the fruit to become over-ripe before picking as this decreases the yield.

Squash, (including marrows) are ready to harvest 6–10 weeks after planting and yield 10–20 tonnes per hectare.

If growing courgettes (Zucchini), harvest them regularly. They should be picked as soon as they reach the correct size, which is not more than 15 cm long.

Sweet melons take 3–5 months to reach maturity and yield 10–15 tonnes per hectare. They are ready to harvest when the shell begins to soften at the blossom end and the fruit comes away from the plant fairly easily.

Water melons yield about 20 tonnes per hectare. They are ready to harvest when the white lower portion starts to go yellow. Another test for ripeness is to tap the melon. If the sound is dull, the melon is ready to harvest, but if a ringing sound is heard the melon is still unripe.

Pumpkins take 3–5 months to mature and can yield 20–30 tonnes per hectare. They are ready to harvest as soon as the skin starts to harden.

Storage

Undamaged pumpkins can be stored for some months in a dry, airy store. Mature squashes (including marrows) can be stored for a short while. They should be hung up in nets in a cool store. Cucumbers and melons cannot be stored.

PESTS AND DISEASES	SYMPTOMS	CONTROL
Melon Fly *Dacus spp* *Figure 41*	The maggots feed inside the fruit causing sunken, discoloured patches, distortion and open cracks. Bacteria and fungi get into the fruit through these cracks and the fruit goes rotten. (Attacks all types of cucurbits, not just melons.) Actual size	*If the pest is present:* 1. Pick all infested fruit and bury it. 2. Mix protein hydrolysate with Fenthion. Spray this on and around the plants. *As a preventive measure:* Spray, starting from flowering, with any of the following: Trichlorphon, Fenthion or Malathion.

PESTS AND DISEASES	SYMPTOMS	CONTROL
Epilachna Beetle (also known as Vegetarian Ladybird or Melon Ladybird) *Epilachna chrysomelina* (see Figure 36 on p. 34)	Both adults and larvae eat the leaves, leaving a network of veins. Damaged leaves shrivel and dry up.	Spray with Malathion, Carbaryl, Trichlorphon, Fenthion or Fenitrothion.
Leaf-footed Plant Bug *Leptoglossus australis*	These bugs puncture the fruit and suck sap from them. Dark spots can be seen on the skin of the fruit. They also attack the shoots which wither and die.	Only spray if the plants are heavily infested. Spray with Gamma BHC or Parathion methyl.

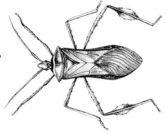

Figure 42   Leaf-footed Plant Bug (life size) Brown with an orange stripe behind the head and with orange and black striped antennae

Tobacco White Fly *Bemisia tabaci* (see Figure 74 on p. 89)	Very small, white insects. They are found mainly under the leaves and fly in a cloud when disturbed. The attacked leaves have yellow patches. These flies can carry viral diseases from one plant to another so it is important to destroy them.	Spray with Fenitrothion Fenthion or Diazinon. If any plants show signs of viral disease (mottling, streaking or curling of the leaves) they should be pulled up and burned before spraying.

PESTS AND DISEASES	SYMPTOMS	CONTROL
Cotton Aphid *Aphis gossypii* **Peach-Potato Aphid** *Myzus persicae*  Actual size *Figure 43*	These pests can become serious during dry weather. They attack shoots and are mainly found on the underside of leaves. Aphids can carry viral diseases from one plant to another. 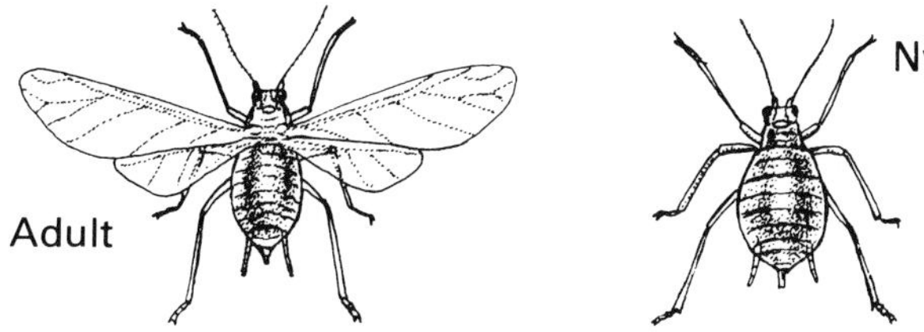 Adult Nymph	Spray with Diazinon, Formothion or Dimethoate.
Powdery Mildew *Erysiphe cichoracearum*	Pale spots on upperside of leaves, and white mildew on the underside. More severe in dry season.	*If plants are infected:* Spray with Benomyl, Mancozeb or sulphur, if infection is not too severe. *As a preventive measure:* Spray with Benomyl, Mancozeb or sulphur.
Downy Mildew *Pseudoperonospora cubensis*	Normally a wet-season disease. Mould develops on underside of leaves.	As for Powdery Mildew.
Stem Rot *Mycosphaerella melonis*	Similar symptoms to Anthracnose.	*If plants are infected:* Spraying with Benomyl will control the spread of the disease. *As a preventive measure:* Spray with Benomyl.
Cucumber Mosaic Virus	Yellow mottling on leaves. (Affects all cucurbits, not just cucumbers.)	There is no chemical control for this disease. Pull up and burn all infected plants as soon as symptoms are seen. *As preventive measures:* 1. Plant resistant or tolerant varieties if possible. 2. Spray Aphids and White Fly to prevent the disease spreading. 3. Disinfect hands after touching infected plants.

PESTS AND DISEASES	SYMPTOMS	CONTROL
Anthracnose *Colletotrichum lagenarium*	The fruit have round, sunken spots with a pinkish-brown centre that later darkens. The leaves have red-brown circular cankers and the stems have elongated, tan-coloured spots.	*If plants are infected:* Remove all ripe and nearly-ripe fruit. Spray with Benomyl. Do not harvest again for at least one week. *As preventive measures:* 1. Buy certified seed. 2. Plant a resistant variety, if available. 3. Dress seed with Captan or Thiram if not already dressed. 4. In wet weather spray routinely with Benomyl. This will only be worthwhile for a valuable crop. 5. Burn crop residues. 6. Practise crop rotation.
Scab or Leak *Cladosporium cucumerinum*	Sunken, dark brown spots on the fruit. There is a gummy ooze from the spots.	This disease cannot be controlled by chemical spraying. Burn all infected plants. *As a preventive measure:* Practise crop rotation.
Fusarium Wilt *Fusarium spp*	Plants wilt suddenly and die.	Burn all infected plants. *As preventive measures:* 1. Plant resistant varieties if possible. 2. Practise crop rotation. 3. Commercial growers could use sterilized soil in the nursery. Seek expert advice on sterilizing soil.

Leeks
Allium porrum

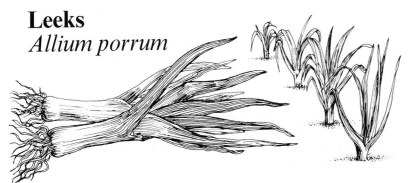

Introduction

The thick white stems of leeks have an oniony flavour and are used as a cooked vegetable, or in soups.

Climatic range

Leeks are tolerant of a wide range of climates and soils, but they prefer cool, moist conditions, and grow best above 500 metres.

Varieties

The following varieties are recommended: Broad Flat, Italian Giant and Musselburgh.

Planting

Sow the seeds at the beginning of the rainy season. Approximately 1.75 kg of seed will be required for each hectare of land to be planted.

Start the seeds off in the nursery. Sow them thinly, 1 cm deep, in rows 15 cm apart. Mulch the seeds, or shade them lightly. Thin the seedlings out if they become crowded.

Transplanting

When the stems of the seedlings are half as thick as a pencil it is time to transplant them. Before planting, trim back the leaves and roots a little as shown in Figure 44. Make planting holes, 15 cm deep and 3 cm wide, using a pointed stick. They should be in rows 35 cm apart, with 20 cm between each plant in the row.

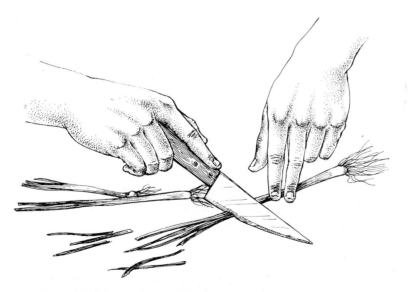

Figure 44   Trim leaves and roots before transplanting

Place the plants in the holes, but do not fill the holes up with earth, (see Figure 45). Just pour a little water into each hole as shown in Figure 46. The space around the plant will fill up with loose soil, and the plant will be able to expand outwards as it grows because the soil is not tightly packed around it.

Figure 45   Place seedling in hole, but do not fill with earth   *Figure 46   Water the seedlings in*

Fertilizer

Apply 200 kg/ha of double superphosphate before transplanting (1 kg tinful for every 140 paces along the row or 1 large spoonful for every 3 paces). Sprinkle the fertilizer along the row and rake it into the soil before making the planting holes.

Three to four weeks after transplanting, top-dress with CAN, at 100 kg/ha (1 large spoonful for every 6 paces along the row).

Harvesting

Leeks should be harvested when they are about 4 cm wide and 25–30 cm long. The crop matures in 4–6 months.

Before sending them to market the leaves and roots should be trimmed and washed.

PESTS AND DISEASES	SYMPTOMS	CONTROL
Onion Thrips *Thrips tabaci* (see Figure 47)	These insects are very small and slender. They cause the leaves to turn silvery and eventually die from the tips downwards.	Spray with Diazinon, Fenthion or Fenitrothion.

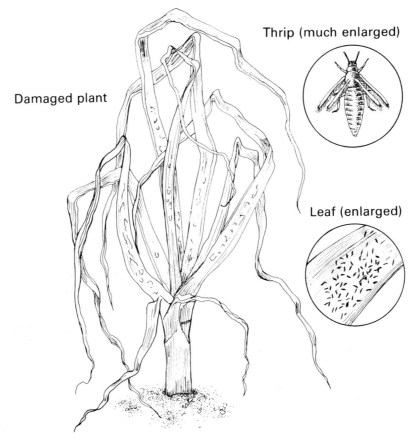

Thrip (much enlarged)

Damaged plant

Leaf (enlarged)

Figure 47   Onion Thrips. The Thrips are hard to see. The leaves become distorted

Lettuce
Lactuca sativa

Introduction

The leaves of lettuce are eaten raw in salads. There is a good market in most large towns for fresh, crisp lettuce.

Climatic range

Lettuce will tolerate a wide range of climates but it requires a good rainfall (over 2,500 mm per year) unless it is grown under irrigation. The soil should be moist at all times while it is growing to ensure that the crop has a good flavour.

At high temperatures, lettuce does not form good hearts and it has a tendency to run to seed.

Soil

Well-cultivated, drained and manured soil is best for growing lettuce, but it will tolerate a wide range of soil conditions.

Varieties

The following varieties are recommended: Great Lakes, Webbs Wonderful and New York.

Planting

Sow thinly in rows 1 cm deep. Provide shade or add a layer of mulch to prevent the soil from drying out. Make small sowings every two weeks to provide a regular supply of fresh lettuce.

When the seedlings have two true leaves, thin them out, leaving 5–10 cm between the seedlings.

Transplanting

When the seedlings are 7 cm tall, transplant them, choosing the strongest seedlings and throwing out the poor ones.

The seedlings should be planted out in rows 30 cm apart, with 30 cm between each plant in the row. Do not plant them too deeply.

Manure and fertilizer

A few days before transplanting, apply some manure and 200 kg/ha of double superphosphate (1 kg tinful for every 170 paces along the row, or one large spoonful for every 3 paces).

After 3 weeks, top dress with 100 kg/ha of CAN or ASN (1 large spoonful for every 6 paces along the row).

Irrigation

Keep the ground moist at all times to ensure even and rapid growth. If this is not done the lettuce may have a bitter taste.

Harvesting

Cut lettuce when the hearts are fully developed. Remove any dirty outside leaves and pack carefully in boxes. Lettuce should be kept cool and sold the same day.

Lettuce is ready to be harvested within 6–12 weeks of sowing, depending on the variety. Yields of 10–14 tonnes per hectare can be obtained.

If lettuces begin to run to seed, pick the whole crop immediately. Discard those that have run to seed, as they will taste bitter, and market the others.

PESTS AND DISEASES	SYMPTOMS	CONTROL
Nematodes (Eelworms) *Meloidogyne spp*	The plants' growth is stunted. The roots are damaged and lumpy.	There is no chemical control for this pest once the crop has been planted. *As preventive measures:* 1. Practise crop rotation. Grass and cereals should be included in the rotation if possible, as nematodes do not attack them. 2. If lettuce is a valuable crop, it may be worthwhile adding a nematicide to the soil before planting.
Heart Rot (a physiological disease)	Lettuces go rotten inside.	Once this disease has appeared it cannot be controlled. *As preventive measures:* 1. The lettuces should be planted farther apart next time. 2. Less nitrogen fertilizer (CAN or ASN) should be used next time.

Maize
Zea mays

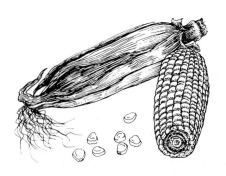 

Introduction

Maize is the staple food in many parts of the tropics. Maize is also an important food for livestock, both the grain and the leaves being eaten. The grain is also used industrially for starch- and oil-extraction.

World production

Brazil	16,309 thousand tonnes per year
Mexico	9,255
Argentina	8,700
Indonesia	5,000
Egypt	2,938
Total world production	394,231

Climatic range

Maize requires an average temperature of 14–30°C; it does not do well at high temperatures and cannot withstand frost. Maize therefore grows well at altitudes up to 2,200 metres. A rainfall of 600–1,200 mm per year is needed and this should be well distributed throughout the growing season. Moisture is particularly important when the seed is setting and a dry period at this time will reduce the yield. Very humid conditions are not suitable for maize and tend to encourage diseases. In very humid areas, planting should be timed so that the cobs will be ripening at the beginning of the dry season. Various types of maize have been developed for different climatic conditions.

Soil

Soils for maize production should be fertile alluvial or loamy soils. They should be free-draining because maize is intolerant of water-logging.

Rotation

Maize takes a lot of the nutrients from the soil, and it can only be grown continuously on very rich, fertile soil, or where large quantities of fertilizer are applied every year. On most soils, it should be grown in a rotation. The rotation should include some type of beans or peas, since these crops return nutrients to the soil.

Seed

Some farmers use their own seed for growing maize, or obtain seed from local sources. Local varieties are usually low-yielding but they have certain advantages: they may be more resistant to some diseases, the cobs are well-sheathed which makes them more resistant to weevil attack in storage, and they may be more acceptable and palatable to local tastes.

Improved varieties give higher yields than local varieties. There are two types of improved variety:
1. Composites
These are stabilized varieties, so the farmer does not need to buy new seed every year. After buying the composite seed in the first year, he can save his own seed for planting in the following years. The yield of composite varieties is usually higher than that of local varieties, but not as high as that of F_1 hybrid varieties.
2. F_1 hybrid varieties
These varieties are very high yielding as long as the fertilizer and rainfall is adequate. They need large quantities of fertilizer every year. The seed should not be saved for planting in the following year. This is because the hybrid variety obtained by crossing two varieties and the seed will not have the same high-yielding characteristics. New seed has to be bought each year.

Planting

Maize should be planted just before the rainy season, or at the beginning of the rainy season. The young maize plant prefers the conditions found at the beginning of the rains. If planting is done later the ground may be too wet for good early growth, and later on the cobs may not fill out because the soil becomes too dry.

A delay in planting may reduce yield considerably. For example, planting two weeks late can reduce the yield by a quarter.

In high-rainfall areas, or on heavy soils, maize can be grown on ridges to improve drainage. Rows should be marked out 75 cm apart (90 cm apart in dry areas). The seeds should be sown 3 cm deep and 25 cm apart in the row (30 cm apart in dry areas). If you are using your own seed and germination is likely to be poor, plant two seeds at each hole and later thin out to leave one plant.

If maize is planted before the rains begin, or if the soil is very light, the seed should be planted more deeply, at 5–8 cm.

Fertilizer

Fertilizer can increase yields considerably, but it will only be worthwhile applying it if the crop is planted early and weeded thoroughly.

Before planting apply 100 kg/ha of double superphosphate and 100 kg/ha of CAN or ASN. Use CAN if the soil is slightly acid, but use ASN if the soil is moderately or strongly acid. (Mix the superphosphate with the CAN or ASN and apply 1 kg tinful for every 50 paces along the row, or 1 large spoonful for each pace along the row.) Mix the fertilizer into the soil well.

The fertilizer can be more effectively used by making a hole beside each seed and putting the fertilizer into this hole. It should be 7 cm away from the seed and 10 cm deep. (Mix the superphosphate with the CAN or ASN and apply one large spoonful for every 4–5 planting holes.)

Six weeks later, when the maize is about knee-high, apply CAN or ASN at a rate of 100 kg/ha (1 kg tinful for every 100 paces along the row, or 1 large spoonful for every 2 paces). The field should be thoroughly weeded before the fertilizer is applied.

A single application of compound fertilizer can be used instead of these fertilizers, particularly where the rainy season is short. Use 20-20-0, and apply 200 kg per hectare, before planting. If the soil is deficient in potash, use 15-15-15.

Weeding

Weeding should begin early, when the plants are about 7 cm tall. The field should be weeded for a second time a few weeks later. Once the maize is about 50 cm tall its leaves will shade the ground and suppress the weeds, so no further weeding will be necessary.

Weeding must be done carefully, as the maize roots grow near the surface and can be damaged by deep hoeing.

In some areas, a parasitic weed known as Striga is a problem. To control Striga, hoe regularly and pull up the plants by hand. Burn the uprooted Striga to prevent it from setting seed. The seeds of Striga can live for up to 15 years in the soil, so it is particularly important to prevent the plants from seeding. Other steps which can be taken to control Striga are as follows:
1. Apply plenty of fertilizer as Striga tends to increase on less fertile land.
2. Practise crop rotation to prevent a build up of Striga in the soil.
3. Apply a herbicide. This will only be worthwhile for commercial growers. Seek local advice on which herbicide to use.

Field hygiene

After harvesting, maize plants should be cut down or ploughed in as soon as possible, to prevent the build up of pests. This will kill about 90% of the Stalk Borer larvae left in the crop. Maize stalks left standing provide a safe home for them until the next crop is planted.

Harvesting and drying

Maize is sometimes cut and stacked up in the field to dry. This allows ploughing or digging to be done in between the piles.

Drying will take 1–3 weeks, depending on the weather. Once dry, the cobs should be picked off the stalks and stored.

The maize stalks can be used for animal bedding or fodder. Alternatively, they can be ploughed or dug in, made into compost or burned. The maize stalks contain half the potash absorbed from the soil by the crop, and almost all the calcium, so ploughing, digging in or making into compost helps to return nutrients to the soil.

Yields of about 6 tonnes per hectare are considered good. Many farmers obtain yields of half this amount or less.

Soil erosion

Maize does not give much protection to the soil, especially if the crop is poorly grown. Planting the seed late, spacing the plants out too widely and not applying enough fertilizer all result in poor ground cover and increase the risk of soil erosion.

Storage

Stored maize can be severely damaged by rats, who feed on the most nutritious part of the maize, the germ. The other major pest is the maize weevil. The build-up of weevils begins with the warm, wet weather at the beginning of the rains. Hybrid varieties of maize suffer more than local varieties, as the tip is not well protected by the sheath and the grain is softer.

Maize stores should be constructed as shown in Figure 19 on p. 14, and the store kept clean as described. When stored on the cob, local varieties of maize will keep for about 4 months, and hybrid varieties for 2 months, without being dusted with insecticide. If the crop is to be stored for longer than this it should be dusted with Malathion, Pirimephos methyl or a Blue Cross insecticide to prevent damage. Stored grain should be dusted with Malathion, Pirimephos methyl, Bromophos or a Red Triangle insecticide. Wash the maize well before cooking. The insecticide will protect the maize for about 8 months. As an alternative to dusting with insecticide, the stored maize can be fumigated with Methyl bromide. This should be done by an expert.

PESTS AND DISEASES	SYMPTOMS	CONTROL
Maize Stalk Borer *Busseola fusca*	Young plants have holes in the leaves. Small blackish caterpillars may be seen in the funnel of the plant. In severe attacks the central leaves may die. In older plants caterpillars may be found boring into the cobs.	*If the pest is present:* Apply Endosulfan or Diazinon granules down the funnel of each plant when they are about 30 cm high. *As preventive measures:* 1. Plough in, dig in, or burn, all crop residues. 2. Observe a closed season of at least 2 months when no maize or sorghum is grown. Plant large areas of maize and sorghum at the same time.

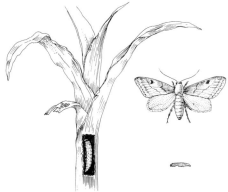 

Figure 48

Pink Stalk Borer *Sesamia calamistes* **Spotted Stalk Borer** *Chilo partellus*	The caterpillars feed inside the stalks, causing stunted plant growth, poorly developed cobs and complete drying up of the plants if severely infested.	*As preventive measures:* 1. Destroy all crop residues and observe a closed season of at least 2 months when no maize or sorghum is grown. 2. Plant large areas of maize and sorghum at the same time. 3. If the pest is likely to be a serious problem, pour Endosulfan or Diazinon granules down the funnel of the plant when it is 30 cm high.

PESTS AND DISEASES	SYMPTOMS	CONTROL
Maize Aphid *Rhopalosiphum maidis*	Soft, dark green insects which feed on the young leaves and flower tassles.	Spray with Dimethoate, Diazinon, Formothion or Fenitrothion.
White Leaf Blight *Helminthosporium turcicum*	Thin, grey oval marks on the leaves.	There is no worthwhile chemical control for this disease. Burn all infected plants. *As a preventive measure:* Plant local varieties as these may have some resistance.
Sorghum Shoot Fly *Atherigona soccata*	The yellowish or whitish maggots bore into the shoots, and the centre of the shoot dies.	*As preventive measures:* 1. Plant the crop early. 2. Observe a closed season of at least 2 months when no maize or sorghum is grown and there are no maize residues in the fields. 3. If the pest is prevalent, spray routinely with Trichlorphon, Fenthion or Fenitrothion. 4. Plant large areas of maize and sorghum at the same time.

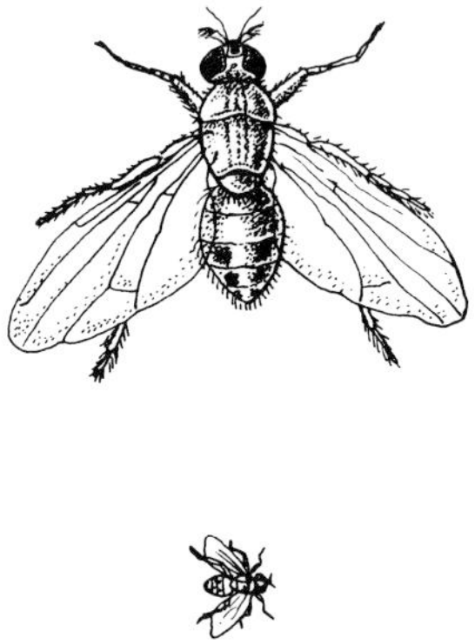

damaged shoot

Actual size

Figure 49

Figure 50

PESTS AND DISEASES	SYMPTOMS	CONTROL

Armyworm
Spodoptera exempta

Leaves eaten or holed down to the midrib.

Spray with Endosulfan or Malathion.

caterpillar
(first pale, then
becomes greenish,
then striped)

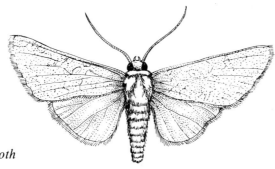

Figure 51   Armyworm moth

Leaf Hopper
Cicadulina mbila
(see Figure 78)

The Leaf Hoppers do not damage the plants much, but they transmit Maize Streak Virus which is a serious disease.

As a preventive measure:
Observe a closed season of at least 2 months, when no maize is grown and there are no maize residues in the fields.

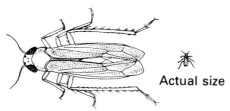

Actual size

Figure 52

Maize Weevil
Sitophilus zeamais

This pest attacks stored maize. It leaves circular holes on the surface of the grain and thin tunnels below the seed coat.

See p. 54, under **Storage**.

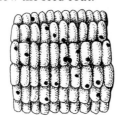

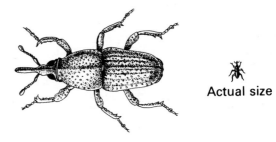

Actual size

Figure 53

57

PESTS AND DISEASES	SYMPTOMS	CONTROL
Maize Streak Virus	Yellow streaks appear on the leaves. The plant is deformed and stunted. This disease is spread by leaf hoppers.	There is no chemical control for viral diseases. Burn all infected plants as soon as they are seen. *As a preventive measure:* Plant the crop as early as possible.

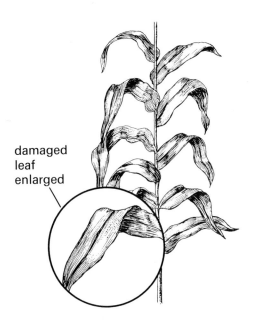

damaged
leaf
enlarged

Figure 54   Maize Streak Virus

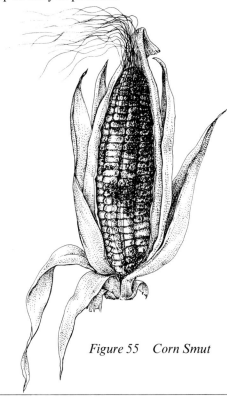

Figure 55   Corn Smut

Corn Smut *Ustilago zeae* (see Figure 55)	Large, blackish lumps appear on the stems, leaves and cobs. These break open to release a fine black powder (the spores which spread the disease to other plants).	There is no worthwhile chemical control for this disease. Pull up and burn all infected plants. *As preventive measures:* 1. Plant certified seed. If using own seed, collect it only from disease-free plants. 2. Practise crop rotation.

PESTS AND DISEASES	SYMPTOMS	CONTROL
Red Flour Beetle *Tribolium castaneum*	These small, reddish-brown beetles are found in flour, and in damaged or broken grains.	Stored grain should be protected using the methods described for Maize Weevil. For maize flour, store only in clean sacks and keep the store clean.

damaged maize seed

Actual size

lava (yellow)

Actual size

Figure 56

Rust *Puccinia sorghi* or *P. polysora*	Red or brown spots on the leaves; the leaves eventually die.	There is no worthwhile chemical control for this disease. All infected plants should be pulled up and burned. *As preventive measures:* 1. Plant resistant varieties if possible. 2. Practise field hygiene.
Downy Mildew *Sclerospora sacchari*	A white mildewy growth appears on the leaves.	As for Rust.

Okra
Hibiscus esculentus

Introduction

Okra is an annual plant with large yellow flowers and slightly hairy pods up to 20 cm long when mature. The immature pods are cooked as a vegetable and sometimes added to soups. They can also be dried. Okra is a useful vegetable to grow in hotter areas. It is often inter-cropped with millet, sorghum, groundnuts or cow peas.

Okra is also known as bhindi or lady's fingers.

Climatic range and soil

Okra requires fairly high temperatures and a well-drained soil but it is tolerant of a wide range of rainfall. Okra grows best below 1,100 metres.

Varieties

The following varieties are recommended: Emerald Green, Perkins Mammoth, Clemson Spineless and White Velvet.

Planting

Okra does not transplant well, so the seeds should be sown direct into the field.

The plants should be in rows 70 cm apart with 30 cm between each plant in the row. Sow the seeds 2 cm deep.

Sow 2–3 seeds at every spot where a plant is required, and when the seedlings are 8–10 cm tall pull up the weaker seedling, leaving only one at each spot.

Fertilizer

Apply 200 kg/ha of double superphosphate before sowing (1 kg tinful for every 70 paces along the row, or 2 large spoonfuls for every 3 paces). Sprinkle the fertilizer along the row and rake into the soil.

Three weeks later, apply CAN or ASN at 100 kg/ha (1 kg tinful for every 140 paces, or 1 small spoonful for every 3 plants).

Harvesting

The pods are ready to harvest 2–3 months after sowing. Pick them when they are less than 10 cm long. They should be harvested regularly to get young, tender pods. Yields of 25–30 tonnes per hectare can be obtained.

Drying

Chop up the pods and spread them out in the sun until dry. Store in clean, dry sacks.

PESTS AND DISEASES	SYMPTOMS	CONTROL
American Bollworm *Heliothis armigera* (see Figure 27 on p. 23)	The caterpillars damage flower buds and pods. The pods may drop before they are ripe.	Spray as soon as young caterpillars are seen, with Fenvalerate, Permethrin, Cypermethrin or Trichlorphon.
Nematodes (Eelworms) *Meloidogyne spp*	The plant's growth is stunted and the roots are damaged.	There is no worthwhile chemical treatment for this pest. *As a preventive measure*: Practise crop rotation. If possible, plant after grass or cereals.
Mosaic Virus	Leaves mottled and distorted.	There is no chemical control for this disease. Burn all infected plants as soon as the symptoms are seen. *As preventive measures*: 1. Plant resistant varieties, if available. 2. If the plants are infested with Aphids or White Fly, spray with Diazinon to prevent the disease from spreading.

Onions
Allium cepa

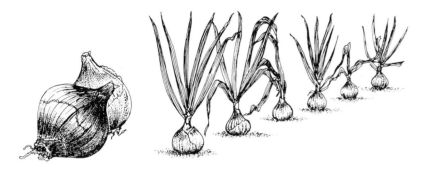

Introduction

Onions are used as vegetables, added to soups and stews, or made into pickles. Fresh onions will store well for several months, and they can also be dehydrated (dried completely) for longer periods of storage.

World production

India	1,600 thousand tonnes per year
Turkey	1,000
Brazil	691
Egypt	536
Pakistan	330
Argentina	300
Colombia	278
Total world production	19,494

Climatic range

Onions require about ten weeks of rain for growth, followed by a fairly long, dry period for ripening. Most varieties prefer relatively cool conditions but some are tolerant of higher temperatures.

Varieties

Red Creole and Tropicana Hybrid (an F_1 hybrid variety) are recommended since they can tolerate high temperatures. White Creole is a good variety for the production of dehydrated onions.

Soil

Onions prefer a firm, sandy loam or alluvial soil. If the soil is acid, applying lime will improve the yield.

Planting

Start the plants off in the nursery. Make raised nursery beds 1 metre wide and of convenient length. Work in manure (about 1 kg per square metre) and double superphosphate at a rate of 200 kg/ha (1 large spoonful per square metre).

Sow the seed in rows 15 cm apart and cover lightly with soil. Shade the beds and water once or twice a day.

When the seedlings are large enough, thin out to about 4–5 cm apart.

Transplanting

The seedlings should be transplanted when they are about half the thickness of a pencil.

The soil should be prepared in advance so that it is well broken down, but firm. Make rows 30 cm apart and apply fertilizer as described below. Plant the onions 8–15 cm apart, depending on the size of the variety.

The onion seedlings should not be planted too deeply; the base of the bulb should only be about 1 cm below the surface. The ground needs to be firm so that the bulbs do not become buried after planting. If the bulbs are completely buried this will reduce their growth.

Direct sowing

Alternatively, onions can be sown direct and not transplanted. If this is done the onion seedlings should be thinned twice. The second thinning should be done when the stems are about the thickness of a pencil, and gives 'spring onions' which can be eaten raw.

Fertilizer

Before transplanting, apply 200 kg/ha of double superphosphate (1 kg tinful for every 170 paces along the row, or 1 large spoonful for every 3 paces). For direct sowing, apply 200 kg/ha of double superphosphate before sowing the seed.

Weeding

Hoe carefully so as not to damage the roots, which lie very near the surface of the soil.

Harvesting

When the onions are fully formed and the leaves are beginning to yellow and dry, bend their stems down. After a few days the stems will wither. Pull the onions up and leave them on the ground to dry for about three days. In very hot conditions, it may be necessary to shade them lightly with dried grass. Turn the onions over every day so that they dry evenly.

Sell or eat very large and thick-necked onions first as these will not keep as well as smaller onions.

Store the onions in slatted boxes, or tie them together and hang up in bunches in a cool, dry place (see Figure 57). Look at the stored onions regularly and remove any that are beginning to rot.

Figure 57   Onions keep well if hung up in a dry well ventilated store

PESTS AND DISEASES	SYMPTOMS	CONTROL
Onion Thrips *Thrips tabaci* (see Figure 47 on p. 50)	The thrips, which are almost too small to be seen, cause silvering and withering of the leaves from the tips downwards. This pest is very common in dry weather.	Spray with Diazinon, Fenthion or Fenitrothion.

PESTS AND DISEASES	SYMPTOMS	CONTROL
Downy Mildew *Peronospora destructor*	The leaves are covered with a brown, downy mildew. Later the leaves turn brown, and eventually they die.	*If plants are infected:* Spray with Benomyl or Metalaxyl. *As preventive measures:* 1. Burn or dig in crop residues. 2. Practise crop rotation.
Purple Blotch *Alternaria porri*	Oval marks, grey with purple centres, appear on the leaves. The leaves curl and die.	*If plants are infected:* Spray with Mancozeb. *As preventive measures:* 1. Burn or dig in crop residues. 2. Practise crop rotation.

Passion fruit
Passiflora spp.

Introduction

Passion fruit are grown for making juice and for fresh fruit; the fresh fruit may be exported. The purple passion fruit, or purple grenadilla (*Passiflora edulis*) is the kind most often grown for juice-making. The yellow-fruited type (*P. edulis var. flaviocarpa*), often known as the yellow grenadilla, is sweeter, and better for the production of fresh fruit. The sweet calabash (*P. maliformis*) is widely grown in the Caribbean islands, and has a grape-like juice. The giant grenadilla (*P. quadrangularis*) is also widely cultivated in the tropics, and there are many other species of *Passiflora* with edible fruit that are cultivated locally.

Climatic range

Passion fruit grow best in warm to cool areas, and therefore prefer altitudes of 1,000–2,000 metres.

A minimum rainfall of 900 mm per year is needed for the growth of this crop, and yields are generally increased by irrigation.

Soil

Passion fruit vines can grow in a wide range of soils, although the most suitable are medium-textured (loamy) soils that are deep and well-drained. The planting site should be sunny and protected from strong winds. The land should be well prepared before planting as the economic life of the vines may be up to five years. The soil should be cultivated as deeply as possible to allow better and faster root growth.

Planting

Passion fruit vines are usually grown from seed, although in some countries cuttings are used.

Passion fruit seeds do not store very well, so they should be planted within three months of being extracted from the fruit. Prepare a seed bed in the nursery about 1 metre wide. Sow the seeds in four drills, 30 cm apart and cover them lightly with soil. A grass mulch, applied after sowing, helps to retain soil moisture and checks weed growth. The mulch should be removed as soon as the seedlings appear.

If the seedlings can be watered, the seeds should be planted about six weeks before the rainy season starts. If water is not available, plant the seeds at the beginning of the rainy season.

When the seedlings are 5 cm high, thin them out to 5 cm apart. After thinning, top dress with 300 kg/ha of CAN (1½ large spoonfuls for every square metre) to stimulate rapid growth.

Transplanting

The seedlings should be transplanted when they are 15–25 cm high; this will probably be about six weeks after sowing, and ideally they should be ready to plant out at the beginning of the rainy season.

The recommended spacing is 2 × 3 metres for unmechanized cultivation; this gives 1,667 plants per hectare. If machines are to be used, then allow at least 3 × 3 metres between each plant; this gives 1,111 plants per hectare.

Planting holes should be prepared at least three weeks before transplanting. Dig holes 45 cm deep and 45 cm across. Keep the topsoil and subsoil separate. Mix the topsoil from each hole with a 20-litre drumful of manure and 125 gm (6 large spoonfuls) of double superphosphate. (This is the equivalent of 140–200 kg double superphosphate per hectare.) Fill the hole with this mixture.

Transplant the seedlings in the evening or early morning. Plant them at the same depth as they were in the nursery and water them to settle the soil around the roots. In hot areas, shade the seedlings after transplanting.

Constructing a trellis

The passion fruit is a vine which naturally climbs up objects such as trees or fences. When cultivated, the vine must be given something to climb up as it grows, and the usual system of support is a trellis.

Construct the trellis using posts that are 2·7 metres long and about 15 cm in diameter. Place these posts in holes 60 cm deep and spaced 6 metres apart in the row. The end posts, in particular, should be firmly anchored in the ground. In each row, stretch a single strand of wire tightly across the top of the posts and secure it to each post with a nail or staple. The whole structure should be strong enough to support the weight of the vines when the fruit are ripe.

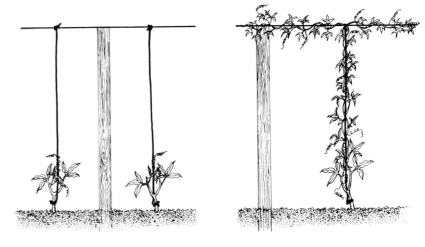

Figure 58   Training passion fruit up a trellis

Training

A piece of twine should be tied around the base of each young plant and the other end tied to the wire above. Select two healthy shoots at the base of each plant and train these up the twine by twisting them around it (see Figure 58).

All other shoots and side branches that emerge should be removed regularly. When the two main shoots reach the wire, train them along it by twisting. Tie them to the wire where necessary.

Pruning

It is important that the side branches hang down freely and do not become tangled up together. To prevent tangling, the tendrils must be removed regularly. Any side branch which trails on the ground should be cut off 15 cm above the ground.

After fruiting, each side branch should be cut back to allow a new side branch to develop. Look for a newly developing side branch as close to the main stem as possible, and cut the old side branch off just beyond this point. Where no new side-branches have formed, cut off the old side branch at the fourth node from the main stem, as shown in Figure 59.

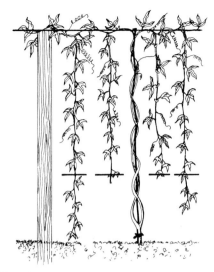

Trimming shoots to 15 cm above ground

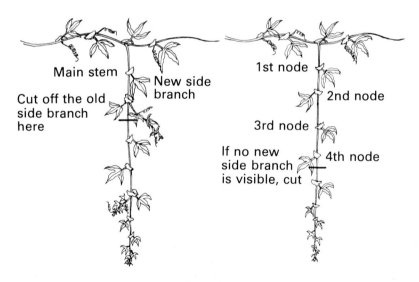

Main stem

Cut off the old
side branch
here

New side
branch

1st node

2nd node

3rd node

If no new
side branch
is visible, cut

4th node

Figure 59   Pruning passion fruit after harvest

Weeding

Keep the field free from weeds. It may be possible to intercrop with a low-growing annual crop, such as beans, especially during the first year, and this will help to control weed growth. But the crop must be harvested, and the ground cleared, before the passion fruit ripen.

Fertilizer

To stimulate growth, apply 120 gm (6 large spoonfuls) of CAN per plant, one month after transplanting.

There is no need for any more fetilizer during the life of the vines.

Irrigation

If the vines are irrigated, the plants will keep growing and flowering for most of the year. This increases the yield and spreads production over a longer season.

Harvesting

The first fruits appear 7–8 months after transplanting, but the main harvest is not obtained until 12–13 months after transplanting. The economic life of well-maintained vines is up to 5 years.

For juice-making, fully mature fruit are required. When the fruit reaches this stage it falls naturally from the vine. It should be collected every morning. For selling as fresh fruit the crop should be harvested in the same way, to obtain the best flavour. But, if necessary, it can be picked from the vine when it is not quite as ripe as this.

PESTS AND DISEASES	SYMPTOMS	CONTROL
Kenya Mealybug *Planococcus kenyae*	Masses of white or pink, soft-bodied insects found on the fruit, stem and leaves. Severe attack by these pests is associated with a sooty mould, and part of the plant, or the whole plant, may wither and die.	Spray with Diazinon, Malathion or Dimethoate.

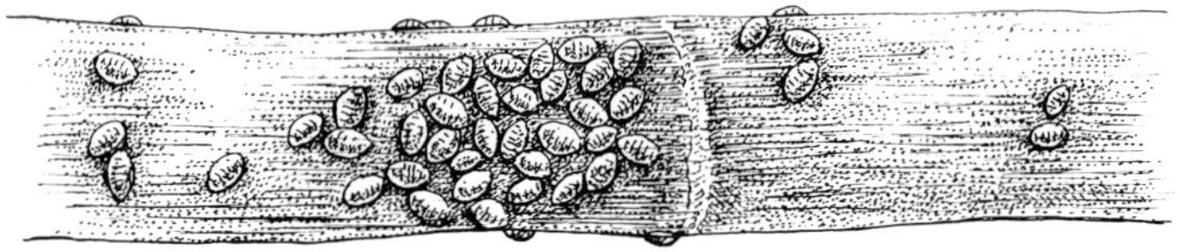

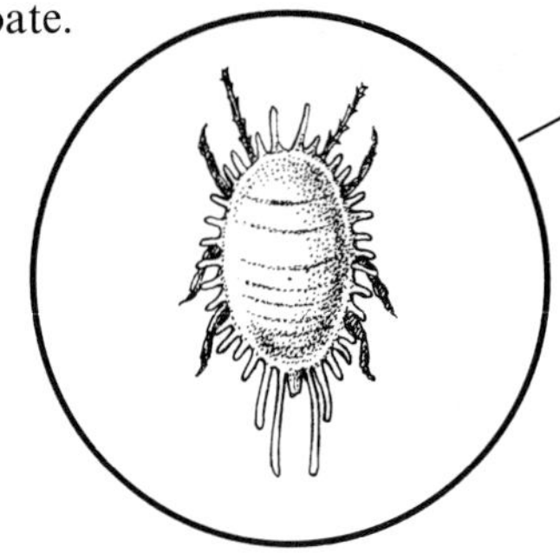

Mealybug (enlarged)

Figure 60

Giant Coreid Bug *Anoplocnemis curvipes* **Leaf-footed Plant Bug** *Leptoglossus membranaceus* (see Figure 42 on p. 46)	These bugs puncture the vines near the terminal buds, which eventually wither and die.	The bugs can be removed and destroyed by hand in small fields where the level of infestation is low. For larger areas, spray with Diazinon, Trichlorphon or Malathion.
Stink Bug *Acrosternum pallidoconspernum*	These bugs puncture the fruit and suck sap from them, causing woodiness. The feeding holes can be seen on the surface of the fruit.	As for Giant Coreid Bug. Harvest as much fruit as possible before spraying.
Aphids (see Figure 90 on p. 113)	Several species of aphid attack passion fruit. Heavy infestation can weaken the vines.	If heavily infested, spray with Dimethoate, Fenitrothion or Diazinon.
Yellow Mites *Polyphagotasonemus latus*	These tiny pests cannot usually be seen but they damage the leaves, flowers and fruit, which may fall or be deformed.	Spray with Dicofol, Chinomethionate, Binapacryl, or sulphur.
Brown Spot *Alternaria passiflorae*	Brown marks on the leaves and fruit. Most severe in humid weather.	For light infections, spray with a copper fungicide. For serious attacks of this disease, spray with Mancozeb or Propineb.

PESTS AND DISEASES	SYMPTOMS	CONTROL
Systates Weevil *Systates pollinosus* (see Figure 89)	The Weevils feed on the edges of the leaves, causing stunted growth in the young plants. They feed at night and rest at the base of the plant during the day.	Spray around the base of the plant with Endosulfan or Fenitrothion.

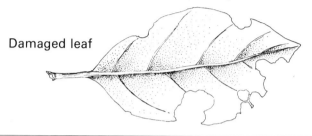

Damaged leaf

Figure 61

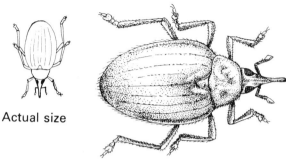

Actual size

Fruit Flies *Ceratitis capitata* and *Dacus spp.* (see Figure 96 on p. 116)	The flies lay their eggs on the fruit, and the white maggots, up to 1 cm long, burrow inside the fruit and feed on the flesh.	Collect and bury all infested fruit. Spray with Fenthion or Fenitrothion, mixed with protein hydrolysate.
Woodiness Virus	This virus disease retards the growth of the plant. The leaves and fruit may become distorted and the affected leaves become discoloured. The fruit produced by the diseased plants are small, hard and have very little juice.	There is no chemical control for this disease. Pull up and burn all infected plants immediately. Wash hands after handling diseased plants and before touching healthy ones, as the virus can be carried on the hands. Wash pruning equipment and other garden tools with soap or other detergent.

Pawpaws (Papaya)
Carica papaya

Introduction

The pawpaw is a short-lived perennial tree, seldom more than 9 metres tall, which is grown for its fruit.

World production

India	250 thousand tonnes per year
Indonesia	220
Brazil	201
Mexico	200
Zaire	174
Philippines	88
Total world production	1,599

Climatic range

A warm to hot climate is needed for this crop, with a rainfall of about 1,000 mm, evenly distributed throughout the year. Pawpaws grow best below 2,000 metres.

Soil

Pawpaw trees need deep, well drained soils which retain moisture. Avoid waterlogged soils.

Male and female trees

There are three types of pawpaw tree, but these can only be distinguished by their flowers:
1. Female trees: these bear female flowers only.
2. Male trees: these bear male flowers only.
3. Hermaphrodite trees: these bear both male and female flowers.

Figure 62   Female and male flowers of pawpaw

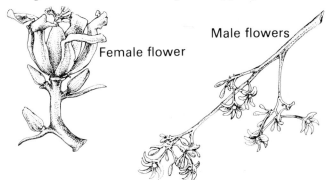

Female flower

Male flowers

The different types of flowers are shown in Figure 62. Female or hermaphrodite trees are required for fruit production. The female trees will need some male trees planted among them, to act as pollinators. Because the different types can only be distinguished after the trees have flowered, it is necessary to grow a number of seedlings at each planting station and then select one plant after the flowers have appeared.

Varieties

Honey Dew	An Indian variety that bears medium-sized fruit.
Kiru	A Tanzanian variety that bears large fruit.
Mountain Pawpaw	A variety that will grow at lower temperatures. It bears small fruits that are generally used for making jams and preserves.
Solo	A Hawaiian variety that bears small, round fruits. All trees are hermaphrodite.

Planting

The seeds can be started off in the nursery or sown direct (see p. 69).

In the nursery, prepare raised seed beds, 1 metre wide and of convenient length. Sow the seeds at a depth of 2 cm in rows 15 cm apart. Water the beds heavily before sowing and continue to water them regularly afterwards. The seeds should germinate within 3 weeks. The seedlings should be transplanted when they are 10–15 cm high.

Transplanting

Make planting holes 60 cm wide and 60 cm deep. The holes should be spaced at 3 metres × 3 metres. Keep the topsoil separate and mix it with a 20-litre drumful of manure and 60 gm (3 large spoonfuls) of double superphosphate. (This is equivalent to 70 kg of double superphosphate per hectare.) Fill the hole with this mixture.

Plant the seedlings in the hole and firm the soil around them. Top dress with 40 gm (2 large spoonfuls) of CAN per planting hole. If you are planting a hermaphrodite variety such as Solo, plant only one seedling in each hole. For other varieties plant four seedlings in each hole and later thin to one plant. This should be done after flowering when you can tell which are the female plants. Pull out most of the male plants, but leave one for every 25–100 females, to serve as a pollinator.

Direct sowing

Space the planting holes at 3 metres × 3 metres and sow six seeds per planting hole. After germination, thin to four plants per hole. After flowering, when it is possible to tell the sex of the plants, thin to one plant per hole. If growing a hermaphrodite variety such as Solo, sow 3 seeds in each hole and thin once only.

Fertilizer

At the beginning of the rainy season each year, apply 200 gm per plant of CAN or ASN (1 kg tinful for every 5 plants). This is equivalent to 220 kg/ha.

Intercropping

Low annual crops like peppers, beans, onions and cabbages may be grown between the rows until the pawpaw trees grow large enough to shade the area. Pawpaws themselves may be planted as an intercrop between young citrus trees.

Harvesting

When a yellow tinge appears on the blossom end, the fruit is beginning to ripen and should be harvested for marketing. For home use the fruit can be harvested when fully ripe. Handle fruit very carefully because it bruises easily.

The yield per tree varies from 30–150 fruits per year, giving a yield of 35–40 tonnes/ha.

The lifespan of a tree can be 10 years, but after the first 3–4 years the yields fall, so new trees should be planted every 4 years. Plant the new trees on a different piece of land.

PESTS AND DISEASES	SYMPTOMS	CONTROL
Systates Weevil *Systates pollinosus* (see Figure 61 on p. 67)	The Weevils eat the edges of the leaves of young plants. They feed at night and rest at the base of the plant during the day.	If the plantation is fairly small, remove and destroy the Weevils by hand. Otherwise, spray the plants and the ground around them with Endosulfan or Fenitrothion.
Red Spider Mite *Tetranychus spp.*	This pest is very small and hard to see. It makes the leaves turn yellowish brown and the fruits become brown and rough.	Spray with Chinomethionate, Dimethoate, or Dicofol. Make sure the undersides of the leaves are well covered by the spray.
Mealybugs *Planococcus spp.* (see Figure 60 on p. 66)	Mealy mass of soft-bodied insects. The leaves are black and sticky with honeydew and there are often ants attending the mealybugs.	Spray a band of Dieldrin or Gamma BHC around the trunk of each tree. This keeps off the attendant ants and allows predators to eat the Mealybugs.
Nematodes (Eelworms)	The roots are misshapen and stunted; tree growth and yield are generally reduced.	*As a preventive measure:* Practise crop rotation. When the trees are replaced, the seedlings should be planted on new land. *If the pest is present:* The soil can be treated with a nematicide, if economically worthwhile, before planting.

PESTS AND DISEASES	SYMPTOMS	CONTROL
Stem Rot *Pythium* and *Phytophthora spp.*	Stem becomes brown and soggy. This disease is caused by a soil-borne fungus and is common in soils with poor drainage.	There is no chemical control for this disease. *As a preventive measure:* Avoid planting in water-logged soils.
Powdery Mildew *Ovulariopsis papayae*	White powdery coating on the leaves.	*If plants are infected:* Spray with Benomyl, if infection is not too severe. *As a preventive measure:* Spray routinely with Benomyl.
Black Rot *Ascochyta caricae*	Circular brown patches on leaves, stems and fruit.	There is no chemical control for this disease. To minimize the spread of the disease, remove all fallen leaves and branches from the plantation. *As a preventive measure:* Remove all leaves, branches and other crop residues from the plantation before replanting.
Black Rust *Asperisporium caricae*	Black spots on the undersides of the leaves.	*As a preventive measure:* Spray with Benomyl, Zineb, Maneb or Mancozeb.
Viral Diseases	The leaves are yellowed, mottled or distorted. Tree growth may be stunted.	All infected trees should be dug up and burned immediately. *As a preventive measure:* For planting, use seeds from trees showing resistance to viral diseases.

Peas
Pisum sativum

Introduction

A climbing plant, whose seeds are eaten boiled. They can also be dried and used to make soups. In some countries, much of the crop is canned or frozen. The mangetout varieties, also known as sugar peas or Chinese peas, are grown for the pods, which are picked when young and eaten whole.

World production

Dry peas

India	430 thousand tonnes per year
Egypt	124
Rwanda	51
Zaire	50
Argentina	47
Burundi	37
Morocco	37
Total world production	12,269

Green peas

India	250 thousand tonnes per year
Mexico	38
Turkey	38
Argentina	36
Ecuador	36
Egypt	36
Total world production	4,699

Climatic range

Plentiful rainfall is required as the soil should be moist throughout the growing season. However, a dry period is needed for ripening. Peas grow best in fairly cool conditions, and are therefore suited to altitudes above 750 metres.

Soils

The soil should be high in organic matter, well-drained and well-cultivated.

Varieties

The length of the rainy season is the most important factor to consider in choosing a variety. Some varieties, such as Earlicrop, mature quickly and are suitable for areas where the rainy season is fairly short. This variety also has the advantage that it does not need staking. The variety Alderman is recommended for areas with a long wet season, while Onward is suitable where the rainy season is of medium length. For hotter areas and dry pea production, the best variety is Black Eyed Susan.

Planting

Dig the soil thoroughly and add compost or manure. Apply 200 kg/ha of double superphosphate or DAP (1 kg tinful for every 70 paces along the double row) before planting and rake into the soil.

60–90 kg of seed is required for every hectare of land to be planted. Sow the seed directly into the field. Plant in double rows 10 cm apart with 60 cm between the double rows. In the row, leave 8 cm between each seed (10 cm for taller varieties). Sow the seed 4 cm deep. Sow a new lot of seed every 3 weeks to give a regular supply of fresh peas.

Weeding and keeping birds off

Keep the ground free of weeds, as peas do not like weed competition, especially when young. It may be necessary to cover the crop with netting as the peas mature, to prevent birds from eating them.

Staking

For the taller climbing varieties, provide twiggy sticks for the plants to climb on.

Harvesting

Green peas can be harvested 8–12 weeks after sowing, but for dry peas the crop must be left until fully mature, which takes about 20 weeks. Green peas must be harvested when young and sent to market quickly. For dry peas, pull up the whole plant and leave them in heaps to dry in the sun before threshing. For details of storage, see p. 13.

PESTS AND DISEASES	SYMPTOMS	CONTROL
American Bollworm *Heliothis armigera* (see Figure 27 on p. 23)	The caterpillars feed on the flower buds, pods and seeds.	Spray as soon as young caterpillars are seen, with Fenvalerate, Permethrin, Cypermethrin, Dichlorvos or Trichlorphon, Profenofos, Quinalphos or Methidathion.
Aphids	Several species of Aphids suck the sap of the plants and cause distortion of the young shoots and leaves.	If heavily infested, spray with Dimethoate, Formothion or Diazinon.
Powdery Mildew *Erysiphe spp*	Powdery white mould covers the leaves and pods. This disease is encouraged by high temperatures.	*If plants are infected:* Spray with Benomyl, Mancozeb, Dinocap or sulphur, if infection is not too severe. *As a preventive measure:* Spray routinely with Benomyl, Mancozeb or sulphur.

Peppers
Capsicum spp

sweet peppers

chilli peppers

Introduction

There are two main types of pepper, chilli peppers and sweet peppers. Chilli peppers are small, red or green in colour, and very hot to the taste. They are used to spice soups, stews and sauces. Sweet peppers are also red or green in colour. They are used as a cooked vegetable, added to stews, or eaten raw in salads.

World production

Chilli peppers

Nigeria	620 thousand tonnes per year
Turkey	480
Mexico	474
Indonesia	200
Egypt	165
Tunisia	128
Total world production	6,671

Climatic range

Peppers tolerate a wide range of climates from warm temperate to tropical, but they do not like very cool conditions and do not grow well at heights of 1,500 metres above sea level or more. In hot, dry areas peppers do well if irrigated.

Varieties

Chilli peppers The following varieties are recommended: Long Cayenne, this variety has green and red fruit with a pungent flavour. Cherry Pepper, the fruit of this variety are small and round with a pungent flavour. Chilli Pepper, a variety with green, purple or red fruits, having a mildly pungent flavour.
Sweet peppers The following varieties are recommended: Emerald Giant, California Wonder and Yolo Wonder.

Do not plant more than one variety of either sweet pepper or chilli pepper. If more than one variety is present, they may cross-pollinate, and poor quality fruits will be produced.

Preventing the spread of disease

Do not plant peppers near tobacco, either in the nursery or the field, as peppers share many diseases with tobacco. Some of these are viral diseases which can be carried on the hands, so wash your hands after handling tobacco plants and before handling peppers. If you use snuff or smoke tobacco, in cigarettes or a pipe, you should always wash you hands before touching the pepper plants.

Planting

Start the seeds off in the nursery. Prepare raised beds and work in some manure and double superphosphate (about 1 large spoonful per square metre). Mark out drills 15 cm apart and sow the seeds thinly, lightly covering them with soil. Give the beds some shade and water once or twice a day.

Later on, thin out the seedlings to 5 cm apart.

Transplanting

Transplant the seedlings when they have four true leaves and are 7–10 cm tall. Plant out at a spacing of 90 cm × 60 cm. Before transplanting, apply manure and fertilizer as described below.

Manure and fertilizer

Before transplanting, add 1–2 handfuls of manure to each planting hole, together with double superphosphate at a rate of 2 small spoonfuls for each hole (250 kg/ha).

When the plants are 15 cm high, apply CAN at a rate of 1 large spoonful for every 4 plants (100 kg/ha). Four weeks later apply more CAN, this time giving 1 large spoonful for every 2 plants (200 kg/ha).

Mulching and cultivation

Apply grass mulch to protect the fruit from water splashes and to retain soil moisture.

When the plant is 30 cm high, the growing point can be pinched out to encourage ripening of the fruit.

Harvesting

The peppers will be ready to harvest 8–12 weeks after transplanting.

Chilli peppers: Chilli peppers should be harvested when mature and turning red, unless green chillis are wanted for the fresh market. To dry chillis, spread them out on mats that have been raised off the ground. At night and when rain is likely, the drying fruit should be covered. The chillis will take several days to dry out. When dry, store in sacks, or threaded on to a string and hung in a well-ventilated place.

After harvesting chilli peppers wash your hands to remove the juice. Avoid getting the juice into your eyes as it will make them sore.

Chilli peppers should be harvested regularly; the harvest continues for about 8 weeks. Yields of 2–3 tonnes/ha are possible.

Sweet peppers Harvest sweet peppers when they have filled out but are still green or just turning red. Do not allow them to become over-ripe as this will reduce the yield. Some varieties can be harvested all at once, but most varieties are harvested at about 5 day intervals over a period of several weeks.

Handle sweet peppers carefully as they bruise easily. Pack and transport to market as soon as possible. Yields of 4–10 tonnes per hectare can be obtained.

PESTS AND DISEASES	SYMPTOMS	CONTROL
American Bollworm *Heliothis armigera* (see Figure 27 on p. 23)	The caterpillars feed on the flower buds and fruit.	Spray as soon as caterpillars are seen, with Fenvalerate, Permethrin, Cypermethrin, Dichlorvos, Trichlorphon, Profenofos, Quinalphos or Methidathion.
Striped Blister Beetle *Epicauta albovittata* (see Figure 37 on p. 34)	The beetles eat irregular holes in the leaves.	Spray with Malathion, Carbaryl or Trichlorphon.
Tobacco White Fly **Bemisia tabaci** (see Figure 74 on p. 89)	Small white insects that rise in a cloud when disturbed. They suck plant sap and may distort the growing points of the plants.	Spray with Diazinon, Fenthion or Fenitrothion.
Jassids *Empoasca spp*	This pest feeds on the leaves, leaving a mosaic of white specks.	As for America Bollworm or Tobacco White Fly.

Figure 63   Jassids. They are light green in colour and move rapidly when disturbed

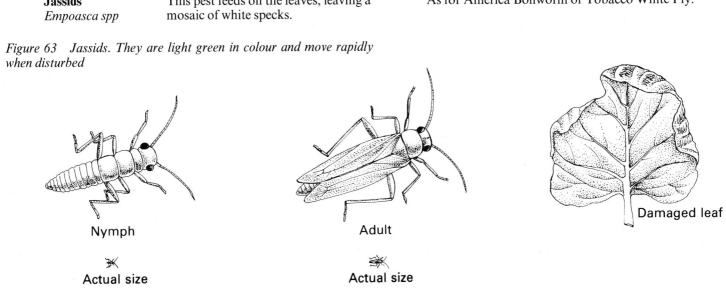

Nymph

Actual size

Adult

Actual size

Damaged leaf

PESTS AND DISEASES	SYMPTOMS	CONTROL
Aphids *Macrosiphum spp* and *Aphis gossypii* (see Figure 90 on p. 113)	Aphids suck plant sap, causing distortion of the plants. They are especially damaging during dry weather.	Spray with Dimethoate, Formothion or Diazinon.
Yellow Tea Mites *Polyphagotarso-nemus latus*	This tiny pest attacks the leaves which become curled and distorted.	If the plants are severely infested spray with Dicofol or Binapacryl.
Powdery Mildew *Laveillula taurica*	This is a common and very serious disease. Yellow spots appear on the upper surfaces of the leaves, and a powdery fungal growth on the undersides of the leaves. Severe loss of leaves also occurs.	*If plants are infected:* Spray with Benomyl, if infection is not too severe. *As a preventive measure:* Spray routinely with Benomyl.
Bacterial Wilt *Erwinia tracheiphila*	The plants wilt and die, even though the soil is moist.	Any infected plant should be pulled up and burned as soon as it is seen. *As preventive measures:* 1. Buy certified seed. 2. Practise crop rotation. Each field should have a rest period of at least 4 years, when no peppers, potatoes, brinjals, tobacco, tomatoes or Cape gooseberries (*Physalia peruviana*, see Figure 64) are grown on it. If the disease has built up to a high level in the soil, the land should not be used for these crops for a period of 10–12 years.

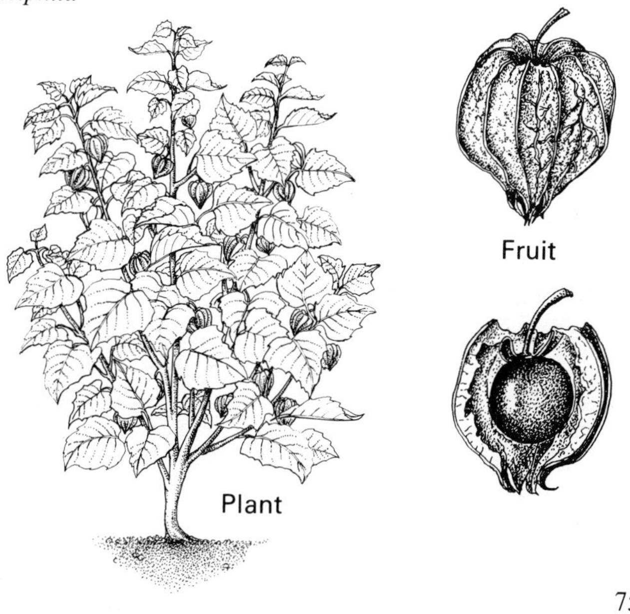

Fruit

Plant

Figure 64 Cape Gooseberry. This is easily grown under medium rain-fall conditions, and has a pleasant fruit

PESTS AND DISEASES	SYMPTOMS	CONTROL
Anthracnose *Colletotrichum capsici*	Grey-brown spots on the fruits, which redden prematurely.	*If plants are infected:* Remove all ripe and nearly-ripe fruit. Spray with Benomyl or Mancozeb. Do not harvest again for at least a week. *As preventive measures:* 1. Buy certified seed. 2. Plant resistant varieties if possible. 3. Dress seed with Captan or Thiram if not already dressed. 4. In wet weather, spray routinely with Benomyl or Mancozeb. This will only be worthwhile for a large and valuable crop. 5. Burn crop residues. 6. Practise crop rotation.
Viral Diseases	The leaves are mottled or streaked; or they may be curled or rolled up. The yield is reduced and if infection is severe the plants may die. These viruses are spread by Aphids and White Fly.	There is no chemical control for viral diseases. If any seedlings show signs of infection they should be burned. For non-commercial growers, infection after the seedling stage is not very serious. *As preventive measures:* 1. Smokers should wash their hands before touching the seedlings. 2. If these diseases are prevalent, plant resistant or tolerant varieties. 3. Spray Aphids and White Fly to prevent the diseases spreading.

Potatoes
Solanum tuberosum

Introduction

The potato is a valuable staple food in many countries. It will grow at higher altitudes where maize does not yield well. For example, in an experiment conducted at an altitude of 2,200 metres, potatoes produced twice as much protein and 25% more starch than maize, and potato growing was more than twice as profitable. Under good management the yields from potatoes are high and the crop can be stored for several months.

World production

Indonesia	10,125 thousand tonnes per year
Jordan	3,400
Chile	2,149
Ecuador	2,066
Uruguay	1,700
Bolivia	1,694
Egypt	977
Total world production	284,471

Climatic range

For good growth, regular rainfall of about 25 mm per week is required. The rainy season should be at least 3 months long, and a longer rainy season will give higher yields. Potatoes prefer cool conditions and therefore grow best above 1,500 metres; they can be grown up to altitudes of about 3,000 metres.

For the production of seed potatoes, potatoes should not be grown below an altitude of 2,300 metres, as the plants must be free of disease, and diseases are more prevalent at lower altitudes.

Soil

Well-drained, light, fertile soils are best. Heavy soils restrict tuber growth.

Varieties

Varieties which have good storage characteristics and resistance to Late Blight disease should be chosen. Some varieties have a three-month growing period while others have a four-month growing period. Those with a longer growing period are higher yielding, but only if there is enough rainfall in the fourth month, so the variety chosen should be suitable for the local climate. Advice should be sought from local seed merchants or research stations on which varieties to grow.

Seed potatoes

Apart from Late Blight, the most serious diseases of potatoes are Bacterial Wilt and viral diseases. These can only be controlled by

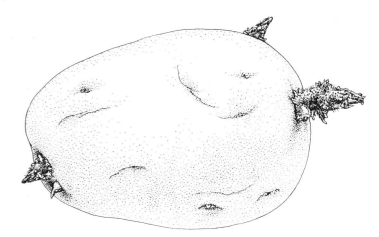

Figure 65 A seed potato with sprouts developing, and ready for planting

using certified seed potatoes. At low altitudes (below 2,300 metres) farmers should buy certified seed potatoes every year. At higher altitudes, where diseases are less prevalent, the farmer need only buy certified seed potatoes once every two years. In alternate years he can save potatoes from his crop to use as seed potatoes.

Depending on the size of the seed potatoes, 2–3 tonnes will be needed for every hectare of land to be planted.

If the seed potatoes can be encouraged to sprout before planting, then none of the growing season will be wasted. They should be placed on trays in a warm, dry shed which has plenty of light but not direct sunlight. Each tuber should develop 4–5 short, sturdy sprouts in about 3–4 weeks, (see Figure 65).

Planting

Planting is usually done at the beginning of the rains to give the potatoes as much growing time as possible.

Potatoes should be planted in ridges since these provide loose soil which allows the tubers to expand. It is also possible to plant potatoes on flat ground, provided they are earthed up later on, but the results will not be as good. (Earthing up is essential as the growing tubers must not be exposed to the light. If they are they will turn green, and green potatoes can be poisonous. Earthing up also protects the tubers from Potato Tuber Moth.) Ridges should be made 75 cm apart at the crest, and the potatoes planted 25–30 cm apart in the ridge.

They should be planted about 10 cm deep in the ridge, but a little deeper than this if grown on the flat.

Growers with ridging equipment can make ridges, and place the fertilizer and potatoes in the furrows. The ridger is then used again, to split the existing ridges and make new ones over the potatoes.

Manure and fertilizer

Manure can greatly increase yields, but because of Black Scurf disease (which sometimes results from contact with manure) it should be mixed in well with the soil before planting. For high yields, 500 kg/ha of DAP (1 kg tinful for every 25 paces along the row) should be applied before planting and mixed into the ridge. On acid soils 200 kg/ha of double superphosphate and 300 kg/ha of CAN should be used instead of DAP. (Mix the two fertilizers together, using 2 tinfuls of double superphosphate to every 3 tinfuls of CAN, then apply the mixture, at a rate of 1 kg tinful for every 25 paces along the row.) Potash is generally not required.

Weeding

Weed the land for the first six weeks; after this the crop's leaves should shade the ground and suppress weed growth.

Harvesting

The potatoes are harvested at the end of the rainy season, or whenever they are ready. If the tops are pulled off two weeks before harvesting then the skins harden and the potatoes are less likely to be damaged. They can be harvested mechanically or by hand. If harvesting by hand use a fork or a forked hoe. Do not push the fork directly into the middle of the ridge as this can damage the potatoes.

Good varieties can yield about 40 tonnes per hectare. In most countries the yield is about 7 tonnes per hectare, but this could be pushed up to an average of 20 tonnes per hectare by improved management, the use of Blight-resistant varieties and a good spraying programme.

Storage of eating potatoes

After harvesting, potatoes will only keep for about two months in

Figure 66 A potato storage shed (for eating potatoes). This should be dark and well ventilated

most parts of the tropics, although in cool conditions they can be kept for much longer. Light spoils potatoes, making them turn green and begin to sprout, so they must be stored in a dark place. Never eat green potatoes as they are poisonous.

For large quantities of potatoes, special stores are worth building. These should be dark inside and have controlled ventilation, (see Figure 66). Ventilation by cool night air is very important as it lowers the storage temperature and keeps the potatoes in good condition for a longer time. Open the vents or the doors at night and close them during the day.

Inside the store the potatoes should be kept in large, strong containers as shown in Figure 66. These should be made of wood and raised 20 cm off the ground to allow air to circulate around the potatoes.

Potatoes can also be stored in a heap on the floor. They will keep better if there is an air vent directly underneath the heap.

If potatoes are to be stored for a long time they can be sprayed with a chemical which stops them from sprouting, such as Propham.

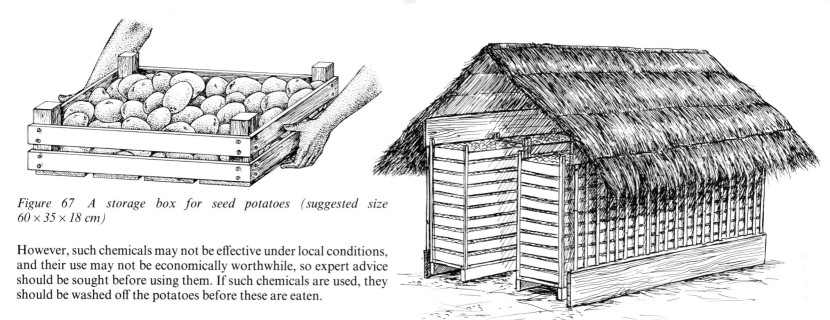

Figure 67 A storage box for seed potatoes (suggested size 60 × 35 × 18 cm)

Figure 68 A storage shed for seed potatoes. This should be light and airy, and the potatoes kept in shallow boxes

However, such chemicals may not be effective under local conditions, and their use may not be economically worthwhile, so expert advice should be sought before using them. If such chemicals are used, they should be washed off the potatoes before these are eaten.

Storage of seed potatoes

If saving seed potatoes from the harvested crop, select small- to medium-sized potatoes that are undamaged and free from disease.

The potatoes required for seed should be obtained from the storage shed, for eating potatoes, about 3–4 months before being needed for planting. They should be transferred to the seed potato store.

Seed potatoes should be stored in the light. Under these conditions the seed potatoes will produce sprouts a few centimetres long, and remain in good condition for planting for six months or more. A good, even light is necessary, but not direct sunlight as this will spoil the seed potatoes. An open-sided store that has a wide, overhanging roof is ideal. Store the seed potatoes in shallow trays (as shown in Figure 68), with the trays separated from each other, so that the light reaches all the potatoes.

Once the potatoes have sprouted they should be sprayed every two weeks with an insecticide such as Dimethoate. This is done to kill aphids which carry viral diseases.

PESTS AND DISEASES	SYMPTOMS	CONTROL

Potato Aphid
Aulacorthum solani
Peach-Potato Aphid
Myzus Persicae and other Aphids

Small soft-bodied insects found underneath the leaves. A sooty black mould develops on the upperside of the leaves. Aphids are a serious pest because they carry viral diseases from one plant to another. These diseases produce cupped, distorted and yellowish leaves.

If any plants show signs of viral diseases pull them up and burn them immediately. Then spray the Aphids with Dimethoate, Diazinon, Formothion or Fenitrothion.

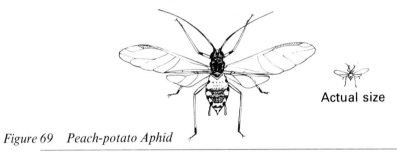
Actual size

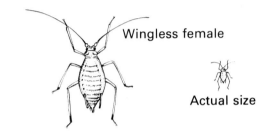
Wingless female
Actual size

Figure 69   Peach-potato Aphid

Potato Tuber Moth
Phthorimea operculella

The moth lays her eggs on potatoes in the field or in the store. The caterpillars burrow into the tubers, making black tunnels.

If the pest is present:
Spray the tubers with Permethrin, Fenitrothion, Malathion, Fenvalerate, Pirimephos methyl, or Cypermethrin before storage.
As preventive measures:
1. Plant the potatoes as deeply as possible and earth up at least twice during the growing season.
2. Harvest potatoes and put them into the store in the morning or early afternoon. The moths emerge to lay their eggs in the early evening, so the potatoes should not be lying about in the open at that time.
3. Keep the tubers in clean, cool stores.
4. If this pest is likely to be a problem, spray the potatoes with Permethrin, Fenitrothion, Malathion, Fenvalerate, Pirimephos methyl, or Cypermethrin before storage.

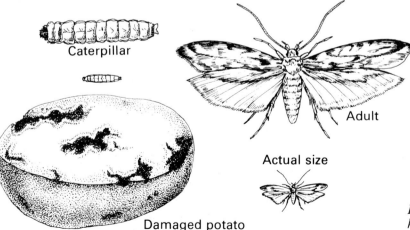
Caterpillar
Adult
Actual size
Damaged potato

Figure 70   Potato Tuber Moth. The adult is grey-brown with whitish hindwings. The caterpillar is pale green

PESTS AND DISEASES	SYMPTOMS	CONTROL
Epilachna Beetle or **Vegetarian Ladybird** *Epilachna spp* (see Figure 36 on p. 34)	The beetles and their larvae feed upon the leaves, leaving a network of veins.	Spray with Fenthion, Fenitrothion, Trichlorphon, Carbaryl or Malathion.
Nematodes (Eelworms) *Meloidogyne spp*	Spots and warty galls on the potato tubers.	*As preventive measures:* 1. Practise crop rotation. The rotation should include grass or cereals if possible. 2. Apply a nematicide such as Carbofuran to the soil before planting. This may not be worthwhile for the small grower.
Late Blight *Phytophthora infestans*	Irregular brown patches appear on the leaves and spread rapidly in damp, cloudy weather. This is a very serious disease which can destroy the whole crop. The disease is only visible after it has got a good hold on the plant, so preventive spraying should start before there are any signs of the disease. Blight thrives in climates where there is no severe dry spell or cold spell. It is spread by air-borne spores from related plants, such as tomatoes.	*As preventive measures:* 1. Plant resistant varieties. 2. Spray routinely with Mancozeb, Propineb or a copper fungicide, starting from emergence. Spray every 2 weeks in dry weather, and once every 4–7 days in wet weather. *If plants are infected:* If more than 5 % of the leaf area is infected then it is too late for preventive spraying. The disease may be eradicated by spraying with Metalaxyl, but this will only be worthwhile for commercial growers.
Bacterial Wilt *Pseudomonas solanacearum*	The plant wilts suddenly and dies. If a tuber from an infected plant is cut open and squeezed, it produces a white juice. This disease attacks other crops of this family, especially tomatoes. The disease builds up in the soil if such crops are grown continuously on the same land.	Any infected plant should be pulled up and burned as soon as it is seen. *As preventive measures:* 1. Buy certified seed potatoes. 2. Plant a resistant variety if possible. 3. Practise crop rotation. Each field should have a rest period of at least 4 years when no potatoes, peppers, brinjals, tomatoes, Cape gooseberries (see Figure 64) or other crops of this family are grown. If the disease has built up to a high level in the soil the land should not be used for these crops for a period of 10–12 years.

PESTS AND DISEASES	SYMPTOMS	CONTROL
Black Scurf *Corticium solani*	Small black lumps develop on the tubers. This disease generally occurs when the potatoes are planted directly next to manure or compost.	*As a preventive measure:* Mix manure or compost into the soil well before planting.
Target Spot or **Early Blight** *Alternaria spp*	Brown, ringed spots develop on the leaves. This disease generally occurs in dry spells when the crop is under irrigation.	*As a preventive measure:* Spray with Propineb, Zineb or Mancozeb, starting from emergence.
Black Leg *Pectobacterium carotovorum*	The plants wilt and the stem becomes a rotting black mass. This disease occurs in very wet seasons.	There is no worthwhile chemical control for this disease. *As a preventive measure:* Buy certified seed potatoes.
Viral Diseases	The leaves become cupped, distorted or yellowish. These diseases are carried in the seed potatoes. They are also carried from one plant to another by Aphids.	There is no chemical control for viral diseases. Pull up and burn any infected plants. *As preventive measures:* 1. Buy certified seed potatoes. 2. Spray Aphids to prevent them from spreading the diseases.

Spinach and spinach beet

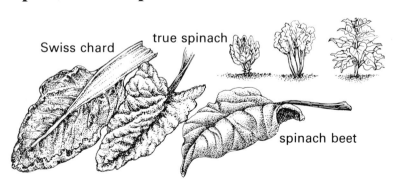

Swiss chard

true spinach

spinach beet

Introduction

These crops are grown for their leaves which are cooked as a vegetable. True spinach (*Spinacea oleracea*) is an annual plant, so it only lasts for one growing season. Other crops which are harvested and cooked in the same way as true spinach, are New Zealand spinach (*Tetragona expansa*) and spinach beet or Swiss chard (*Beta vulgaris*). These are often referred to as 'spinach', but in fact they are entirely different plants from true spinach. One important difference is that New Zealand spinach and spinach beet are biennial plants, so that the crop lasts for two years.

Many other spinach-type crops, of local origin, are grown in different parts of the tropics.

Climatic range

True spinach needs quite a lot of moisture, about 2,500 mm per year, and cannot withstand very high temperatures. It grows best at altitudes of 2,000 m or more. New Zealand spinach needs less moisture and can be grown in areas that are too dry for true spinach. It does well below 1,500 metres. Spinach beet will tolerate a wide range of climatic conditions as long as the soil is well-drained.

Soil

These crops need a well-drained, fertile soil that is high in organic matter.

Varieties

Monstrous Viroflay is a good variety of true spinach. Recommended varieties of spinach beet are Fordbrook and Lucullus.

Planting

Sow the seed directly into the field. True spinach seed needs to be soaked in water before it is planted to improve its germination. Put the seed into cold water the day before you intend to plant.

For true spinach, sow the seed thinly, in rows 40 cm apart. After a few weeks, thin out the seedlings leaving about 15 cm between each plant.

Because New Zealand spinach and spinach beet last for two years they need more room. They should be sown thinly, in rows 90 cm apart. After a few weeks, thin out the seedlings, leaving about 25 cm between each plant.

The seedlings that are pulled up when thinning can be cooked and eaten.

Cultivation

Regular weeding is needed. In dry weather the crop should be watered regularly.

The growing point of New Zealand spinach can be removed, at first harvest, to increase leaf production.

Fertilizer

Before sowing apply double superphosphate and rake into the soil. (*For true spinach*, 1 kg tinful for every 125 paces along the row, or 1 large spoonful for every 2–3 paces.)

After 3 weeks, top dress with 100 kg/ha of CAN. 1 large spoonful for every 5 paces.

For New Zealand spinach and spinach beet, use twice the quantity.

Harvesting

Pick leaves when required. Do not pick all the leaves of a plant at once. If selling the crop, take it to the market as soon as possible.

PEST	SYMPTOM	CONTROL
Slugs	Holes eaten in the leaves.	Scatter slug pellets around the plants. Or you can bury a tin in the ground, with its rim level with the soil surface. Fill the tin with beer. The slugs are attracted by the smell of the beer and drown.

Sweet potatoes
Ipomoea batatas

Introduction

Sweet potatoes are an important food crop, especially in lower altitude areas. The tubers are boiled or roasted and the young leaves are also eaten as a vegetable. Sweet potato vines are a useful fodder crop, especially in the dry season. Some varieties are especially suitable for this, producing a large number of leaves.

World production

Vietnam	2,400 thousand tonnes per year
Indonesia	2,350
India	1,545
Brazil	1,516
Korean Republic	1,387
Philippines	1,037
Burundi	943
Total world production	113,954

Climatic range

Sweet potatoes grow best in areas receiving 750 mm of rain per year or more, but they are also drought resistant, remaining green and healthy even during severe droughts. They do well in both warm and cool conditions and so may be grown at altitudes up to 2,100 metres. They can even grow as high as 2,400 metres.

Soil

Sweet potatoes will thrive in a wide range of soils, but for good yields the soil should be fertile. Heavy or stony soils will not allow the tubers to grow well.

Varieties

There are many different varieties. Some have hard yellow tubers, while others have sweet, soft tubers, usually with white flesh.

Planting

The land must be prepared well in advance of planting. Sweet potatoes can be grown on ridges or mounds (see Figure 76 on p. 93); provide a looser soil and allow the tubers to expand more easily. The ridges should be made 90–150 cm apart.

Planting can be done at any time when the soil is fairly moist.

Cuttings should be taken from the central stem of the vine, from leaf stalks or from tubers. Pieces of stem and stalk should be 30–40 cm long. They are buried in the ground at a 45° angle, with about two-thirds of the cutting in the soil (see Figure 71). Tubers, or parts of tubers, are buried completely. The lower leaves are removed from stem cuttings before planting. The cuttings should only be taken from disease-free plants.

The cuttings are planted 30–60 cm apart along the row or ridge. The exact spacing is not very important.

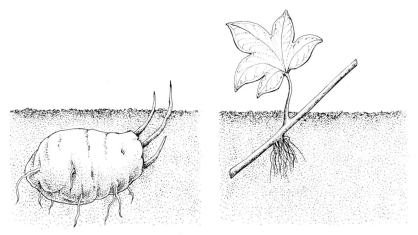

Figure 71   Planting tubers and cuttings of sweet potatoes

Manure and fertilizer

Nitrogen fertilizers are not recommended when sweet potatoes are grown for their tubers, because they result in more leaves being produced, and fewer tubers. But manure or compost increase yields considerably. They should be added to the soil before planting the cuttings, at a rate of about 50–100 kg/ha.

Weeding

Weed between the plants until their leaves cover the ground. After this weeding will not be necessary.

Harvesting

The first tubers are ready for harvesting 4–6 months after planting. Sweet potatoes are usually only harvested as required, since they cannot be stored fresh for more than a few days. The largest tubers are selected for harvesting; they can be found by looking for cracks in the ground. The tubers are dug up with a sharpened stick or fork, not with a hoe.

Sometimes the whole crop is harvested at once and dried for storage. The tubers are cut into chips before being dried.

Yields of 10–20 tonnes per hectare should be obtained.

PESTS AND DISEASES	SYMPTOMS	CONTROL
Sweet Potato Weevil *Cylas puncticollis*	The adult weevils feed on the leaves and the soft parts of the vines; the larvae tunnel into the tubers and stems.	*If the pest is present:* Spray the leaves with Fenthion, Endosulfan or Gamma BHC, to kill the adult weevils. *As preventive measures:* 1. After harvesting, burn all infested plants. 2. Make sure there are no weevils or larvae on the cuttings before planting. 3. Plant new cuttings as far away as possible from the previous season's crop.

Actual size

Figure 72

PESTS AND DISEASES	SYMPTOMS	CONTROL
White Fly *Bemisia spp.* (see Figure 74 on p. 89)	Small white flies that rest under the leaves and rise in a cloud when disturbed. They carry viral diseases from one plant to another.	Spray with Fenthion, Fenitrothion or Diazinon. Before spraying, pull up and burn all plants showing signs of viral disease.

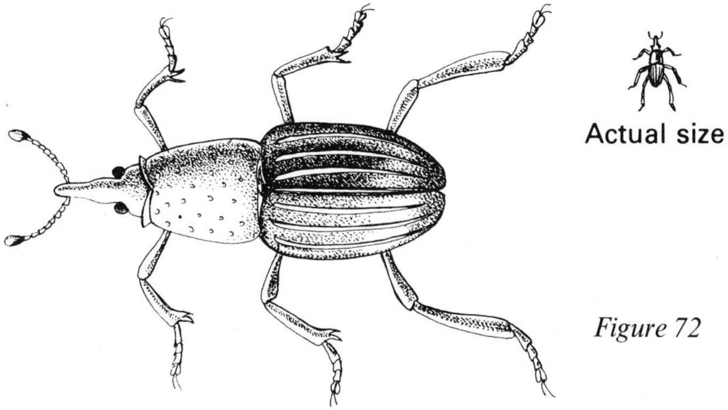

PESTS AND DISEASES	SYMPTOMS	CONTROL
Sweet Potato Moth *Omphisa anastomosalis*	The caterpillars of this moth bore into the main stem of the vine.	Spray with Gamma BHC.
Sweet Potato Virus B	The vines are stunted and yellow, with more side branches than normal. This disease is spread by White Fly.	There is no chemical control for viral diseases. Pull up and burn all infected plants as soon as they are seen. *As preventive measures:* 1. Only take cuttings from disease-free plants. 2. Plant resistant varieties if possible. 3. Spray White Fly to prevent them from spreading the disease.
Black Rot *Ceratocystis fimbriata*	Black spots appear on the tubers which eventually go rotten.	*As preventive measures:* 1. Practise crop rotation. 2. Only take cuttings from disease-free plants.

Tomatoes
Lycopersicon esculentum

plum tomatoes

Introduction

This crop is grown for its fruit, which are cooked as vegetables, eaten raw in salads or used in chutneys. Good fruit that have been carefully handled command high prices.

World production

Turkey	3,136 thousand tonnes per year
Egypt	2,431
Brazil	1,500
Mexico	1,082
India	730
Argentina	570
Total world production	49,201

Climatic range

Tomato plants grow well in warm climates and are fairly adaptable to different conditions, but excessive humidity and very high temperatures reduce the yield. Under very wet conditions tomatoes suffer badly from disease and the fruit fails to ripen.

Varieties

Some tomato varieties have round fruit, while others have elongated fruit. The elongated types are known as plum tomatoes, and these varieties are usually grown for canning, although they can also be sold fresh.

Tomatoes generally need staking, but there are some bush varieties which do not need stakes. Some tomato varieties are F_1 hybrids, and these are generally higher yielding, with better resistance to disease. (If growing F_1 hybrid varieties, remember that the seed should not be kept for planting in the next season, as it will not give good plants.)

The following varieties are recommended for canning: Roma, San Marzano, Rutgers 10X Hybrid, Heinz 1350.

The following varieties are recommended for the fresh market: Money Maker, Early Beauty, Super Marmande, Ponderosa, Hotset, Best of All, Marglobe, Alicante.

Sowing

Start the seeds off in the nursery. Choose a nursery site where potatoes, brinjals, peppers, tobacco or Cape gooseberries (*Physalis peruviana*, see Figure 55), have not been grown in the last three years, to reduce the risk of disease. Do not use manure on the nursery bed as this too can spread disease to tomato seedlings.

Sow the seed in drills, 20 cm apart and 1 cm deep.

Later on, thin the seedlings out, leaving 7 cm between each plant.

Transplanting

Transplanting can take place about one month after sowing, or when the plants are 10–15 cm high.

Crop rotation is particularly important for tomatoes, and the area in which they are planted should not have had potatoes, brinjals, peppers, tobacco or Cape gooseberries growing on it for at least 3 years. Diseases can also be transmitted to tomatoes from tobacco, so if you smoke or use snuff you should wash your hands before touching the seedlings.

When transplanting, select only the strong, healthy seedlings. Space the plants at 100 cm × 50 cm. Before transplanting apply fertilizer as described below. Plant the seedlings so that the lowest leaves are just above the surface of the soil. In high rainfall areas tomatoes can be grown on mounds or ridges. Transplanting should be done in the evening, or on a cloudy day. Try to transplant with some soil still attached to the roots of the seedling. Water them in well.

Manure and fertilizer

If the soil is poor in organic matter, apply 2–3 kg of manure per square metre before transplanting, or add 1 kg to each planting hole.

Before transplanting, apply 200 kg/ha of double superphosphate (1 large spoonful to every 2 planting holes). The fertilizer should be mixed in well with the soil.

When the plants are 25 cm high, top dress with CAN at 100 kg/ha, (1 kg for every 100 paces along the row or 1 large spoonful for every 4 plants).

Four weeks later, apply CAN again if possible. Use 200 kg/ha, or 1 large spoonful for every 2 plants.

Mulching and wind protection

Mulch with chopped grass to keep the soil temperature down and retain water in the soil.

Tomato plants should be sheltered from too much wind by fences.

Weeding and watering

Weed regularly to reduce weed competition and diseases. In dry weather, regular watering is essential. The ground should be given a good soaking 2–3 times each week. Avoid splashing water onto the leaves of the plants, as this can encourage the spread of diseases, particularly Blight. To make the best use of the water available and reduce the risk of speading disease, a pot can be sunk into the ground next to each plant, as shown in Figure 1 on p. 2, and the water poured into this pot.

Pruning

Leave only one main stem. Side stems should be pinched out as they grow, every week. When 4–6 trusses of fruit have formed, pinch out the growing tip so that no more fruit can set (see Figure 73). This will encourage the growth of good-sized tomatoes. (If you smoke remember to wash your hands first.)

As the leaves age, remove the ones that are close to the ground, to prevent the entry of Blight.

Staking

Push a 2 metre stake firmly into the ground next to each tomato plant. Tie the stem loosely to the stake as the plant grows. A trellis can be used instead of individual stakes. To construct the trellis, put a strong pole into the ground every 4 metres along the row. Run two wires along the row, at the top and bottom of these poles. For each plant, run a string from the bottom wire up to the top wire. As the plants grow, twist them carefully around the strings.

Harvesting

For canning, the tomatoes should be ripe, but for the fresh market they should be slightly under-ripe so that they travel well. The fruit should be graded for size and colour to get the best prices. Fruit for market should be packed in strong containers to prevent damage.

Yields of 20–100 tonnes/ha can be obtained.

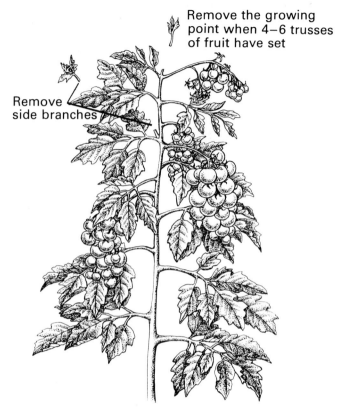

Remove the growing point when 4–6 trusses of fruit have set

Remove side branches

Figure 73   Pinch out side stems (or shoots) each week

PESTS AND DISEASES	SYMPTOMS	CONTROL
American Bollworm *Heliothis armigera* (see Figure 27 on p. 23)	The caterpillars bore into the fruit and feed on the inner parts. Fungi enter through the holes and cause the fruit to rot.	Spray as soon as young caterpillars are seen, with Fenvalerate, Permethrin, Cypermethrin, Dichlorvos or Trichlorphon.
Red Spider Mites *Tetranychus cinnabarinus, T. lombardini* and *T. telarius*	These tiny pests are mainly found underneath the leaves. The leaves become yellow and spotted.	Spray with Binapacryl or Dicofol.

PESTS AND DISEASES	SYMPTOMS	CONTROL
Tobacco White Fly *Bemisia tabaci*	These small white flies are found under the leaves and rise in a cloud when disturbed. They suck plant sap, and may carry viral diseases.	Spray with Fenitrothion, Fenthion or Diazinon. If any plants show signs of virus infection (curled, streaked or mottled leaves) they should be pulled up and burned before the plants are sprayed.

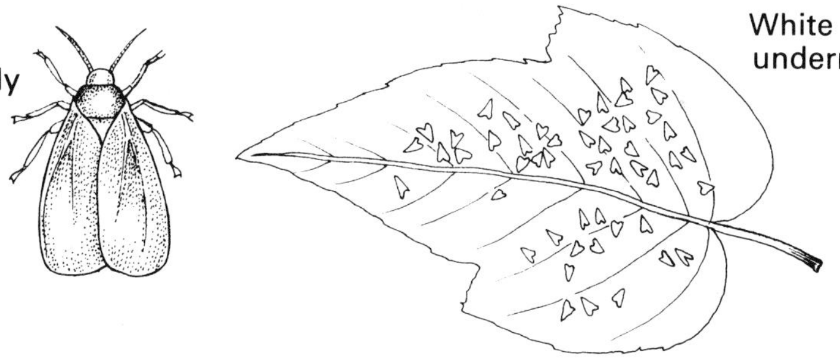

Single White Fly (enlarged)

White Flies resting underneath a leaf

Figure 74

Tomato Russet Mites *Aculops lycopersicei*	These tiny pests cause the tomato stalks to turn a reddish-brown colour at ground level, and this colour then spreads upwards. This pest is more common in dry weather.	Spray with Binapacryl or sulphur. After harvesting, burn all infested plants.
Late Blight *Phytophthora infestans*	Mainly occurs in wet weather, with rapid withering and drying out of the leaves, and brownish, dry rot of the fruit. The whole crop may be destroyed. The disease is only visible after it has got a good hold on the plants, so preventive spraying should start before there are any signs of the disease. This disease also attacks potatoes.	*As a preventive measure:* Spray with Mancozeb or Propineb. Start spraying when the seedlings emerge and spray once every 2 weeks in dry weather, but once every 4–7 days in wet weather. *If plants are infected:* If more than 5 % of the leaf area is infected, then it is too late for preventive spraying. The disease can be eradicated by spraying with Metalaxyl, but this will only be worthwhile for commercial growers.
Early Blight *Alternaria solani*	The effects of this disease are similar to those of Late Blight. Stem cankers develop on the seedlings and brown spots appear on the leaves. The leaves and fruit drop off.	As for Late Blight, but Metalaxyl is not effective against Early Blight.

PESTS AND DISEASES	SYMPTOMS	CONTROL
Nematodes (Eelworms) *Meloidogyne spp.*	Nematodes damage the roots, causing wilting and stunted growth.	*If the pest is present:* There is no effective chemical control. Do not plant tomatoes on the same land for at least 5 years. In the following season, it is best to plant cereals or grass. *As preventive measures:* 1. Practise crop rotation. It is best to plant tomatoes after cereals or grass. 2. Choose a nursery site on ground that has not recently been used for tomatoes, potatoes, tobacco, okra or yams. 3. If nematodes are prevalent, use a granular nematicide (such as Phenamiphos, Dazomet or Carbofuran) in the nursery, before sowing seeds.
Bacterial Canker *Corynebacterium michiganense*	The leaves wilt and the stems may split open. This disease is carried in the seeds and is only visible after it has got a good hold on the plants.	*If plants are infected:* 1. If any diseased seedlings come up in the nursery, destroy *all* the seedlings by burning. Do not use this area again for tomatoes. 2. Burn all infected plants. *As preventive measures:* 1. Buy certified seed. 2. Do not use manure in the nursery as it may contain diseased tomato seeds. 3. Practise crop rotation. 4. If the disease is already apparent, or if it is prevalent locally, disinfect hands and knife when pruning. This should be done between every 2 plants. Dip your fingers and knife into a disinfectant, such as Dettol.
Bacterial Wilt *Pseudomonas solanacearum*	Plants wilt badly, even when soil is moist.	There is no chemical control for this disease. Any infected plant should be pulled up and burned as soon as it is seen. *As preventive measures:* 1. Buy certified seed. 2. Practise crop rotation. Each field should have a rest period of at least 4 years when no crops of this family are grown on it.

PESTS AND DISEASES	SYMPTOMS	CONTROL
Blossom End Rot (a physiological disease)	A water-soaked spot develops on the fruit. It turns brown and enlarges to cover half the fruit. 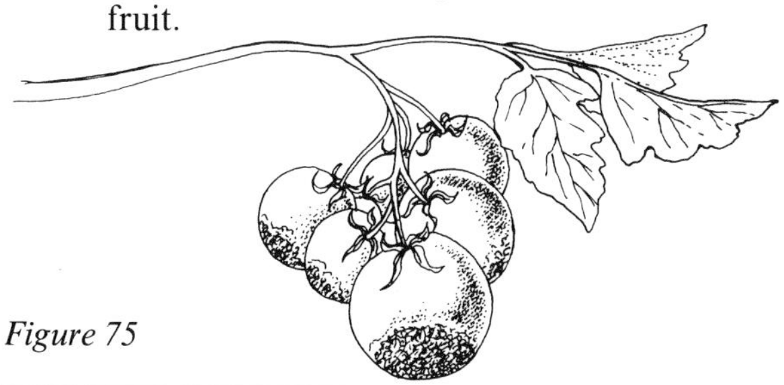	This disease is caused by: 1. Irregular or infrequent watering. 2. Too much nitrogen in the early stages. 3. Calcium deficiency in the young fruit. *If the disease appears:* Regular watering may make it go away. *As preventive measures:* 1. Water regularly. 2. Apply less nitrogen fertilizer (CAN).

Figure 75

Leaf Spot *Septoria spp.*	Small brown spots on the lower leaves, spreading to all leaves after a time. This disease is especially prevalent in wet weather.	*If plants are infected:* Spray with a copper fungicide or Mancozeb before infection is too severe, repeating the spray every week. *As a preventive measure:* Spray once a week with a copper fungicide or Mancozeb.
Leaf Mould *Cladosporium spp.*	Mould develops on the underside of the leaves, which eventually shrivel and turn brown.	*If plants are infected:* Spray with a copper fungicide.
Viral Diseases	The leaves are mottled, streaked or curled. The yield is reduced and if infection is serious the plants may die. These viruses are spread by Aphids and White Fly.	There is no chemical control for viral diseases. If any seedlings show signs of infection they should be burned. For non-commercial growers, infection after the seedling stage is not usually serious. *As preventive measures:* 1. Smokers should wash their hands before touching the seedlings. 2. If viral diseases are prevalent, resistant or tolerant varieties should be grown. 3. Spray Aphids and White Fly to prevent the diseases spreading. 4. For a valuable crop, disinfect hands and pruning knife between plants when pruning, as for Bacterial Canker.

Yams
Dioscorea spp.

Introduction

This popular tubular crop is the staple food of many West African countries and is also widely grown in the Caribbean islands and in Southeast Asia. There are eight different species of cultivated yam, of which four are widely grown. The white Guinea yam (*Dioscorea rotundata*) is the most popular type in West Africa, since it is the best for making 'fufu' (pounded and steamed yam). In Southeast Asia the most commonly grown type is the water yam (also called the greater yam or winged yam, *D. alata*). The flesh has a more watery texture than that of the white Guinea yam and the tubers mature in a shorter time. The two other widely grown species are the yellow yam (*D. cayenensis*) which has yellow flesh and takes longer to mature than the white Guinea yam, and the lesser yam or Chinese yam (*D. esculenta*), which has white tubers with a much sweeter taste than the other species. The lesser yam is often higher yielding than other yams, but it is not widely grown in West Africa as it cannot be used in many of the traditional dishes.

Climatic range

Tropical areas which have a long rainy season are best for this crop. Most types of yam need at least six months of rain, and the average rainfall per year should be 1,500 mm or more. Temperatures of 25–30°C or more are best for growing yams. Below 20°C the plant's growth is stunted and yams cannot tolerate frost.

Soil

A good, fertile soil is needed for yams, and they are often planted as the first crop when new land is opened to cultivation. A loose soil is needed since hard or heavy soils prevent the expansion of the tuber and make harvesting difficult. The soil must be well-drained as yams cannot tolerate water-logging.

Rotation and intercropping

Crop rotation should be practised to maintain soil fertility and avoid the build-up of pests such as nematodes. Yams are often rotated with maize and cassava. Yams are almost always intercropped with crops such as okra, pumpkins, melons or peppers.

Preparing planting material

Yams can be grown from cuttings, but the most common planting material is tubers. Small, whole tubers are best, but pieces of tuber can also be used. About 2 tonnes per hectare are required.

When cutting a tuber for planting, take care not to infect it with bacteria which may cause rotting. If a rotten tuber is cut, throw it out and kill the bacteria on the knife by heating it in a flame or dipping it in a disinfectant such as Dettol for a few minutes. After cutting, dip all the pieces of tuber into wood ash, Benomyl or Captan, to prevent attack by fungi in the soil. If treating the tubers for Yam Beetle, this should also be done at this stage – see p. 95.

Tuber planting material can be pre-sprouted by burying it in a shallow ditch about three months before planting time. After $2\frac{1}{2}$ months the tubers should be dug up carefully and stored in a shady, well-ventilated place until planting time.

Planting

Planting should take place at the beginning, or just before the beginning of the rainy season.

Yams are generally grown on ridges or mounds, as shown in Figure 76. The land should be dug or ploughed well and manure and fertilizer mixed into the soil, before the mounds or ridges are made. The mounds are usually about 50 cm high (knee-high) and spaced about 1–1·5 metres apart. Ridges are spaced about 1 metre apart.

The tubers are planted about 1 metre apart in the ridges. If grown on mounds, then 3–4 tubers are planted on each mound. The tubers are planted about 10 cm deep.

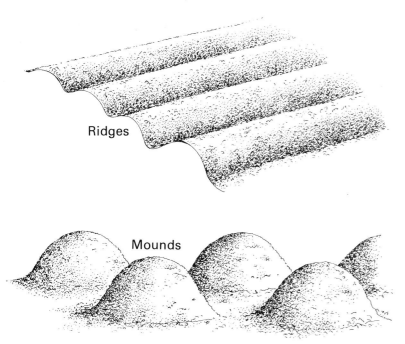

Figure 76   Two different methods of land preparation for growing yams, cassava and other root crops

Fertilizer

Apply sulphate of ammonia before planting, at a rate of 150–250 kg/ha (one large spoonful for each tuber or mound). Where the soil is deficient in potash, a compound fertilizer, 12-12-18, can be used instead, at a rate of 600 kg/ha (3 large spoonfuls for each tuber or mound). Manure or compost should also be added to the soil if available.

Mulching and weeding

Yams are very sensitive to high soil temperatures and dryness. They should be covered with mulch immediately after planting, or yields will be greatly reduced. Weeding is essential for the first 2–3 months, as the young plants are sensitive to weed competition. Earth up around the plants while weeding.

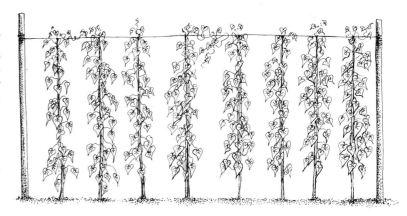

Figure 77   Different methods of staking yams

Staking

When the vines are 1 metre long, about a month after the shoot first emerges, they will need staking. Bamboo poles about 2 metres long are generally used for this. There are several ways of staking the vines, as shown in Figure 77.

Harvesting

There are two different ways of harvesting: single harvest and double harvest.

Single harvest: Once the tubers are mature, the whole plant is harvested. The tubers are mature when the leaves on the vine turn yellow, and the harvest can take place at any time after this, but not more than two months afterwards.

Double harvest: The yams are harvested in two stages. The first harvest is taken 4–5 months after the vine emerges.

For the first harvest, carefully loosen the tuber from the soil, taking care not to damage the roots. Cut the tuber from the plant just below the crown (see Figure 78) and replace the earth around the roots.

The second harvest is taken in the normal way when the plant is mature. Second harvest tubers are oddly shaped and rather unsuitable for eating, but they do make good planting material and they are generally used for this.

Yields of yam vary between 6 and 25 tonnes per hectare.

Storage

The usual way to store the tubers is to tie them up in a yam barn. The sides of a yam barn are a series of upright poles, and the roof is usually made of thatch. The yams are tied to the upright poles with string, as shown in Figure 79. The yams must be completely shaded from the sun, and there must be good ventilation to prevent them from rotting. Inspect the stored yams regularly and remove any rotten ones.

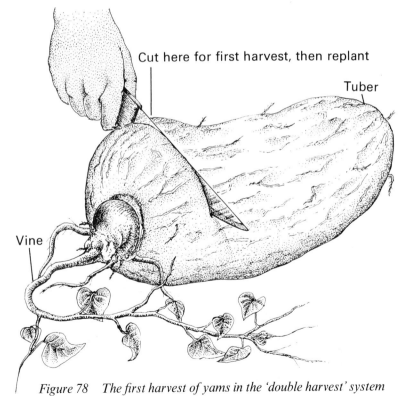

Cut here for first harvest, then replant

Tuber

Vine

Figure 78   The first harvest of yams in the 'double harvest' system

Figure 79   Yams stored in a barn

PESTS AND DISEASES	SYMPTOMS	CONTROL
Yam Beetle *Prionoryctes spp.* and *Heteroligus spp.*	The beetles feed on the tubers in the soil. 	If this pest is prevalent, dust the tubers with Aldrin or Gamma BHC before planting. 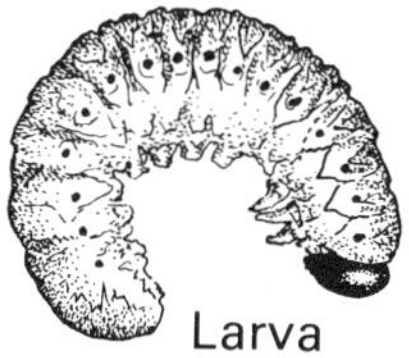

Figure 80   Yam Beetle (actual size)

Adult

Larva

Other beetles	The larvae feed on the leaves and young shoots.	Pick off the larvae by hand. If the vines are seriously infested, spray with Carbaryl
Scale Insects (see Figure 93 on p. 114 for typical Scale Insects)	Small, flat insects with no visible legs attached firmly to the leaves.	Spray with Malathion or Gamma BHC.
Nematodes (Eelworms) *Sculellonema bradys* and *Meloidogyne spp.*	Nematodes make the seed tubers turn brown and prevent them from sprouting. They also attack growing tubers making them look warty.	The soil can be treated with a nematicide but this may not be economically worthwhile. The nematicide is mixed into the soil before planting or dusted onto the tubers before they are planted. *As preventive measures:* 1. Practise crop rotation. The rotation should include grass or cereals (such as maize) if possible, and should leave each plot free of yams for at least four years. 2. Make sure that tubers for planting are free of nematodes.
Mosaic Virus	The leaves are mottled, the growth of the vine is stunted and many extra side shoots appear.	There is no chemical control for viral diseases. Pull up and burn all infected plants. *As a preventive measure:* Only take planting material from disease-free plants.

Section 3
Orchard crops

Apples

Malus pumila

Sharps Early	Will grow at altitudes of 2,500–3,000 metres.
Blenheim Orange	Will grow at altitudes of 2,300–3,000 metres. This variety is high yielding, with large green fruit.
Jonathan	Will grow at altitudes of 2,300–3,000 metres. This variety has good quality fruit and is recommended as a pollinator.
Winter Banana	Will grow at altitudes of 2,000–3,000 metres. A very vigorous variety with large, attractive fruit. Recommended for apple production, and as a pollinator.
Rome Beauty	Will grow at altitudes of 2,000–2,500 metres. This variety requires a vigorous rootstock.
Bramley's Seedling	Will grow at altitudes of 2,500–3,000 metres. This variety produces very good fruit; it requires a pollinator. A cooking variety
King of Thomkin County	Will grow at altitudes of 2,000–3,000 metres. This variety has large, good-quality fruit but is low yielding. A cooking variety

Introduction

There are two types of apples: dessert apples and cooking apples. Dessert apples do not need to be cooked before eating. Some varieties will keep for many months if carefully handled and stored.

Climatic Range

Apples are not a tropical crop, but they may be grown in those areas where high altitude produces a cooler climate, with fairly low temperatures for part of the year.

Varieties

Many apple varieties need trees of another variety, planted alongside them, to act as a pollinator. Seek local advice on which varieties to plant and whether a pollinator is required.

There are many good varieties and below are listed a few as examples. Your local research station should be able to advise you.

Grafting

The best way of growing apples is to plant grafted trees, as these give high yields of good-quality fruit. Buds from high-yielding trees are grafted on young trees which are disease resistant. These grafted trees can be obtained from government or commercial nurseries.

Spacing

The spacing of the trees depends on the variety. For small varieties, such as Sharps Early, a spacing of 2 metres × 2 metres should be used. For medium-sized trees, such as Winter Banana, a spacing of 2·5 metres × 2.5 metres should be used. For large trees, such as Bramley's Seedling, a spacing of 3 metres × 3 metres should be used.

Land preparation

Tree rows should follow the contour on steep slopes (see Figure 4 on p. 3).

Make holes, 60 cm wide and 60 cm deep, keeping the topsoil and subsoil separate. Mix the topsoil from each hole with 1–2 20-litre drumfuls of manure or compost, and 60 gm (d large spoonfuls) of double superphosphate. Use this mixture to refill the hole.

Use the subsoil to make an irrigation basin around the tree, (see Figure 1 on p. 2).

Planting

Loose-rooted trees (those grown in nursery beds) should be transplanted during the cooler part of the year, when they are dormant. Young trees that have been grown in pots or bags can be transplanted at any time.

Plant the tree in the middle of the hole and ensure that the roots spread naturally. Firm the soil in around the tree and water. For those varieties that require a pollinator, a row of the pollinator variety should be planted after every 2–3 rows of the main variety.

Intercropping

A newly planted orchard can be intercropped with low-growing vegetables. These plants will not compete with the trees, and intercropping helps to offset the investment cost of planting the orchard.

Pruning

The purpose of pruning during the first three years is to establish a cup-shaped framework of strong, main branches.

In the first year

Immediately after planting, give the tree its first pruning. If it is an unpruned, one-year-old tree, cut it back to half its height, leaving three or four good buds, as shown in Figure 81.

In the second year

Select three or four strong branches to form the future main branches of the tree. These branches must be at least 15 cm apart and the

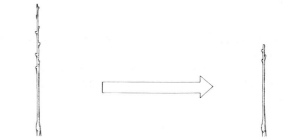

Year 1: immediately after planting, cut back leaving 3–4 buds

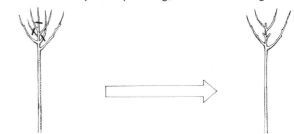

Year 2: select 3–4 branches and remove the rest

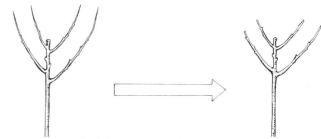

Year 3: at the end of the growing season, cut back the branches

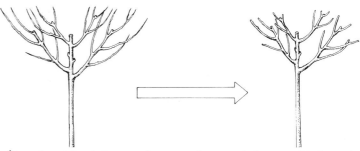

Year 4: cut back the previous year's growth by one third to a half

Figure 81   Pruning an apple tree

lowest branch should be about 1 metre from the ground. Remove all other branches.

Towards the end of the growing season, or during the dormant period, cut off the tips of the main branches to stimulate the growth of side branches.

In the third year

If the tree is growing strongly, cut back the previous summer's growth by a third of its length. If the growth is thin and weak, cut back by half its length.

After the third year the basic shape of the tree should be established. From then on, pruning should be done only to remove dead, broken or criss-crossing branches.

Manure and fertilizer

If available, apply several forkfulls of manure or compost to each tree every year.

Phosphate and nitrogen fertilizers should also be applied every year. The amount to apply for each tree is shown in the table below.

	Double superphosphate	CAN
Year 1	150 gm	500 gm
Year 2	300 gm	700 gm
Year 3	450 gm	800 gm
Year 4	600 gm	900 gm
Year 5	750 gm	1.0 kg
Year 6	900 gm	1.2 kg per tree
Year 7	1.0 kg	1.3 kg
Year 8	1.2 kg	1.4 kg
Year 9	1.4 kg	1.5 kg
Year 10	1.5 kg	1.5 kg
Year 11	1.7 kg	1.5 kg
Year 12	1.8 kg	1.5 kg
Year 13	2 kg	1.5 kg
Year 14	2.1 kg	1.5 kg
Year 15	2.2 kg	1.5 kg
	Double superphosphate	CAN
Year 16 and all following years	2.5 kg	1.5 kg

In the first year, apply the fertilizer three months after planting. In the following years, apply the fertilizer at the beginning of the rainy season.

All manure, compost and fertilizer should be worked lightly into the soil under the tree canopy (the area shaded by its leaves).

Fruit thinning

Normally, some of the flowers and young fruit drop off naturally, so that the right number of fruit are obtained. But sometimes too many apples set and remain on the tree, and the apples are then very small due to overcrowding. If too many apples do set, thinning of the fruit is necessary to ensure large, good-quality fruit. Pick off the small fruit, leaving only two or three evenly spaced apples in each cluster.

Harvesting

When picking apples, handle them carefully. Hold the fruit in the palm of the hand (not with the fingers) and lift upwards. Place the fruit gently into a basket or a container made of soft material. Bruised or damaged fruit fetch lower prices and will not store for long.

Apples can be stored in a cool, dry, rat-proof room or shed. They can be stored in boxes, but will keep better if each apple is individually wrapped in newspaper or tissue paper before being put into the box. This prevents rot spreading from one apple to another.

Growers producing quality apples for sale will need to have spraying equipment and chemicals ready, so that spraying can be done early on in a pest attack. Growers should inspect their trees at least twice a week to scout for pests and diseases.

Local research stations will be able to advise on the main pests and diseases in the area.

A pre-season spray is an effective first control and this should be done when the buds and new leaves are just about to open. Tar oil should be sprayed on the trees because this kills lichens and insect pests, before they can build up in number.

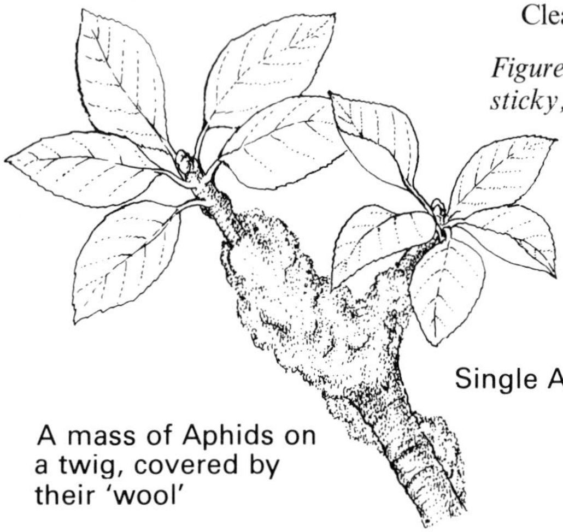

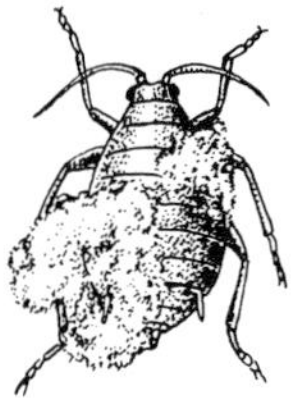

PESTS AND DISEASES	SYMPTOMS	CONTROL
Woolly Aphids *Eriosoma spp*	Woolly Aphids suck plant sap. They spread Canker from one tree to another.	Spray with Malathion, Diazinon, Dimethoate or Formothion. Clean out old cankers where the aphids can hide.

Figure 82   Woolly Aphids. The Aphids are covered with a sticky, white 'wool'

Single Aphid (enlarged)

A mass of Aphids on a twig, covered by their 'wool'

Red Spider Mites *Tetranychus spp*	These tiny, red-brown pests are found mainly underneath the leaves. The upper surface of the leaves may turn a rusty yellow colour and the lower surface may be silvered.	Spray with Dicofol, Binapacryl or Chinomethionate. Repeat the spray one week later.
San José Scale *Quadraspidiotus perniciosus*	A layer of tiny, flat, scale-like insects, that have no visible legs and do not move about, can just be seen on the tree trunk, leaves and fruit. The bark of the trunk may crack and ooze gum. Eventually the tree dies.	Spray with tar oil or Parathion methyl during the dormant period.

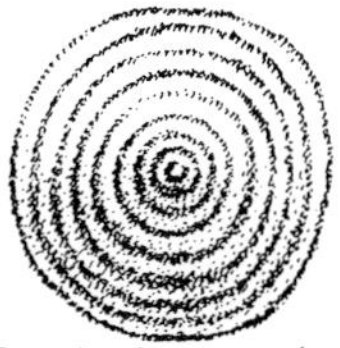

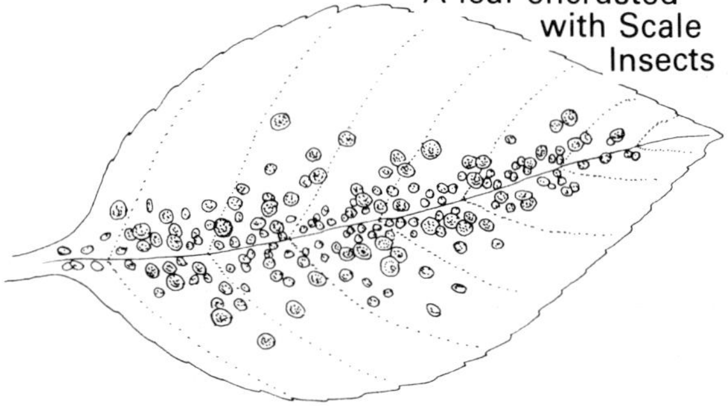

A leaf encrusted with Scale Insects

Figure 83

Single Scale Insect (enlarged)

PESTS AND DISEASES	SYMPTOMS	CONTROL
Apple Scab *Venturia inaequalis*	Fungal disease that attacks leaves and fruit. Leaf margins turn brown and dark spots of irregular shape, appear on the fruit.	*As a preventive measure:* If the disease is prevalent, spray with a copper fungicide, Mancozeb, Thiram or Captan, once every 2 weeks, starting from the pre-blossom stage and continuing until just before harvest.
Apple Mildew *Podosphaera leucotrica*	The leaves and young stems are coated with a fine white or grey powder; they may be slightly distorted.	After harvesting, cut out any badly affected shoots. *As a preventive measure:* If the disease is prevalent, spray with Dinocap or sulphur at 10-day intervals. Begin when the buds are just about to open.
Apple Canker *Nectria galligena*	Sunken areas of dead bark on twigs and young stems. Eventually these expand to encircle the stem which then dies.	Cut off infected twigs and small branches and burn them. If larger branches or the trunk are infected, cut away the infected bark and all discoloured wood, and burn this too. When all the canker has been cut away, paint the area underneath with a mixture of tar oil and copper oxychloride. *If the disease is severe:* Spray with a copper fungicide when the leaves are falling, and again when the buds are about to open.
Root Rot *Armillaria mellea*	A layer of white fungus develops under the bark of the roots and trunk. The tree slowly dies.	Dig out infected trees and burn them. Remove all leaves, roots and dead wood from the hole and burn. Leave the hole open for at least one year to kill the fungus.

Avocados
Persea americana

Introduction

The avocado is very nutritious, with a high protein and oil content. It is a valuable export crop.

World production

Mexico	373 thousand tonnes per year
Brazil	141
Dominican Republic	137
Indonesia	65
Haiti	53
Ecuador	35
Zaire	24
Total world production	1,384

Climatic range and soil

Avocados grow well in areas with a rainfall of 1,000 mm per year or more, and where temperatures are not too high. Altitudes of 1,500–2,000 metres are best for this crop.

The soil should be light, deep and well-drained.

Varieties

The following varieties are recommended:

Fuerte	This variety has fruits of a very good flavour, with thin, pebbly skins.
Haas	A variety which is easy to grow. It has black fruits and is very vigorous.
Nabal	The fruit of this variety have a good flavour, but it only bears fruit once every two years.
Puebla	This variety has round, deep-purple fruit. It is often used as a rootstock.

Some varieties are best when grafted on to a rootstock of another variety. You should seek local advice on whether grafting is needed. If you are skilled at grafting, you can grow the avocados from seed and then graft them. Alternatively, you can buy young trees that have already been grafted.

Growing from seed

Select healthy, egg-sized seeds and plant them in boxes or seedbeds. As soon as they have germinated, transplant the seedlings into pots or tins, about 15 cm in diameter.

Planting

Plant out the young trees at the beginning of the rainy season. Space the trees at 9 metres × 9 metres. Dig holes 60 cm deep and 60 cm wide, keeping the topsoil and subsoil separate. Mix the topsoil with two 20-litre drumfuls of manure or compost and 120 gm (6 large spoonfuls) of double superphosphate.

Remove the plant from its pot, keeping as much soil around the roots as possible. Place in the hole and use the topsoil mixture to fill the hole up. Water in immediately. Use the subsoil to make an irrigation ring around the tree (see Figure 1 on p. 2).

If young leaves are coming out on the trees provide some shade.

Wind protection

On exposed sites, some protection against the wind will be needed. If avocados are exposed to strong winds, they lean to one side and the leaves and fruit may drop off.

Manure and fertilizer

If available, apply manure at the beginning of each rainy season. Fertilizer should also be applied at this time. The best way to decide how much to apply is to have the soil tested by a soil-testing station or laboratory. If this is not possible then you should use the amounts of the different fertilizers and manure for each tree, each year. See table on p. 103.

Irrigation

Irrigation during dry periods will improve the yield. Dry weather makes the fruit fall off but the trees generally survive.

Minor nutrients

Minor nutrients may be in short supply in avocado orchards. Discoloured and mottled leaves are the usual sign of a deficiency. In many countries there are agricultural laboratories which can analyse a leaf sample and tell the farmer which minor nutrient is deficient. If this service is not available, the table below may enable you to identify which nutrient is lacking in your soil.

AGE OF TREE (YEARS)	MANURE	DOUBLE SUPER-PHOSPHATE	CAN	MURIATE of POTASH
1–3	1 20-litre drumful	200 gm	100 gm	none
4–5	1 20-litre drumful	450 gm	200 gm	none
6–7	1½ 20-litre drumfuls	650 gm	450 gm	200 gm
8–9	1½ 20-litre drumfuls	650 gm	650 gm	450 gm
10–14	none	1·0 kg	1·0 kg	650 gm
15 years or more	none	1·2 kg	1·3 kg	650 gm

Table: Fertilizer and manure needs of Avocadoes

MINOR NUTRIENT	SYMPTOMS AND TREATMENT
Zinc (Zn)	The leaves are mottled with light yellow areas between the veins. The growing shoots are unusually large. Apply zinc sulphate, using 250 gm for a 1-year-old tree, 500 gm for a 2-year-old tree, 750 gm for a 3-year-old tree and so on, up to a maximum of 4·5 kg. Scatter the zinc sulphate in a circle beneath the outer edge of the canopy, so that water dripping off the leaves washes it into the soil.
Manganese (Mn)	The leaves lose their green colour. Spray the young leaves with a solution of manganous sulphate.
Iron (Fe)	The leaves lose their green colour. Apply an iron chelate, using 350 gm for a 1-year-old tree, 700 gm for a 2-year-old tree, 1,050 gm for a 3-year-old tree, 1,400 gm for a 4-year-old tree and so on. Scatter on the ground in the same way as described for zinc.

If there are high concentrations of chlorine (Cl) in the soil, the tips of the leaves will turn brown. At the same time, the chlorine may damage the roots of the trees. To compensate for the chlorine, apply DAP fertilizer or a compound containing calcium (*eg* lime) or magnesium.

Pruning

In the first 4–5 years the trees can be pruned to give them an open, spreading branch system. After this, pruning is only done to remove dead branches.

Harvesting

It is not easy to tell when avocados are ready for harvesting, especially in the varieties where the colour of the fruit does not change with maturity. Harvest a few fruits and keep them at room temperature. If they soften within ten days, without shrivelling, then the fruits of that age are ready for harvesting. If they do shrivel they are not yet ready.

PESTS AND DISEASES	SYMPTOMS	CONTROL
Purple Scale Insect *Chrysomphalus aonidum*	Small, flattened, scale-like insects with no visible legs (see Figure 86 on p. 107), that are found attached to the leaves and fruit.	Spray with Diazinon and White Oil, Fenitrothion and White Oil, or Malathion and White Oil.
Fire Ant *Solenopsis geminata*	The ants bite through the bark to feed on the sap. They may also damage the shoots, flowers and fruit. These ants have a very painful bite, and although they do not harm the trees very much, they are a serious pest because they attack people working among the trees.	The ants nest in the soil near the trees. Spray the nests and the leaves of the trees with Gamma BHC or Diazinon.

Figure 84   *Fire Ants. They are a dark reddish-brown and give a painful bite*

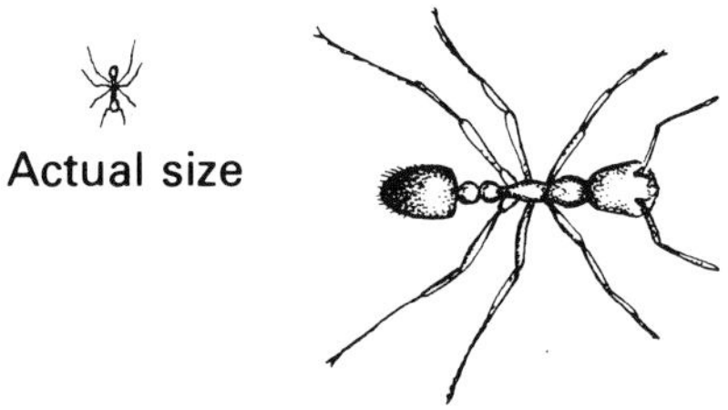

Actual size

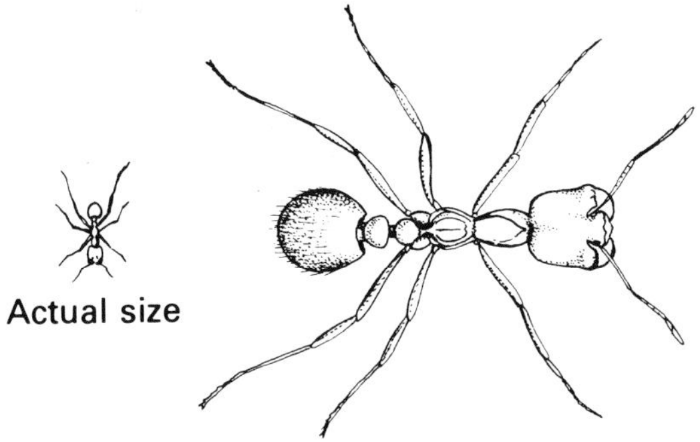

Actual size

PESTS AND DISEASES	SYMPTOMS	CONTROL
False Codling Moth *Cryptophebia leucotreata* (see Figure 94 on p. 115)	The moth lays its eggs on the fruit and when the caterpillars hatch out they bore into the pulp.	Collect all infested fruit from the ground, pick all infested fruit from the tree, and bury it in the ground at least 50 cm deep. *As a preventive measure:* Where this pest is prevalent, spray with Fenthion or Malathion once a week, starting when the fruit are small.
Leaf Spot *Cercospora spp*	Brown spots appear on the leaves.	If trees are severely infected, spray with a copper fungicide.
Fungal diseases (Various species)	The fruit become rotten.	*As preventive measures:* 1. Practise field hygiene and remove all dead wood from the trees regularly. 2. Handle fruit carefully when harvesting.

PESTS AND DISEASES	SYMPTOMS	CONTROL
Root Rot *Phytophthora cinnamoni*	The roots of the tree become rotton and the tree eventually dies.	*As preventive measures:* 1. Do not plant avocados on badly drained soil, as water-logging encourages this disease. 2. Practise field hygiene and remove all dead wood from the trees regularly. 3. Only plant healthy seedlings.

Cashews
Anacardium occidentale

cashew nut   cashew apple

Introduction

The cashew is grown for its edible kernel, which is eaten as a nut or used in confectionery. The cashew apple can also be eaten. The shells of cashew nuts are used in the manufacture of paints and plastics. Deshelling cashew nuts is difficult and is best done in a factory.

Cashew nuts are a useful crop to grow in low-rainfall areas, and have the advantage of needing no fertilizer. They are sometimes grown in a mixed orchard with mangoes, bananas, coconuts or citrus.

World production

India	150 thousand tonnes per year
Mozambique	150
Brazil	80
Kenya	58
Tanzania	58
Philippines	42
Total world production	509

Climatic range

Once they are established, cashew trees need very little water, and can be grown in areas with as little as 500 mm of rain per year. They do best in hot climates and do not grow well above an altitude of 500 metres; they cannot be grown at all above 1,000 metres.

Soil

The soil should be deep and well-drained. Cashews will grow quite well on infertile sands, but not on coral outcrops.

Seed

Seed for establishing new trees should be selected from healthy, high-yielding trees. Before planting, sort out the good nuts by the following method:
1. Take a small bucket and fill it with sea water, or water to which salt has been added.
2. Put the cashew nuts in the bucket of salt-water. Some will float and others will sink. Those that float will have poor germination and growth so throw them away. Only plant the seeds that sank to the bottom of the bucket.

The nuts for planting should be dried in the sun for several weeks before they are stored or planted. Do not store seed for more than one year before planting.

Planting

The seeds should be sown at the beginning of the rainy season. (In

Point where nut was
attached to cashew apple.

Plant this way up.

Figure 85   Cashew seed

areas where there are short and long rainy seasons, plant them at the beginning of the long rains.)

Clear the field of weeds and dig holes 2 metres across and 75 cm deep. There is no need to add any manure, compost or fertilizer to the planting holes. Refill each hole with the topsoil.

The spacing of the holes should normally be 12 metres × 12 metres, but in very dry areas the trees should be spaced farther apart than this.

Plant 3 seeds per hole, burying them 6–8 cm deep. Leave 20 cm between each seed. Place the seed so that the stalk points upwards as shown in Figure 85.

At the centre of each planting site, drive in a strong stake so that the seedling can be supported.

Intercropping

During the first few years, low-growing vegetables, or crops such as cotton, sesame, groundnuts cowpeas or grams may be grown in the field (The cultivation of these crops is described in *Tropical Field Crops* volume 3 in this series.) These crops should not be planted too close to the cashew trees: allow at least 2 metres between the trees and the intercrop.

Management

The seeds should germinate within 10–12 days. If the soil is dry the seedlings should be watered.

Protect the seedlings from animals by placing wire cages or thorny branches around them.

After 3–4 months, thin to leave only one plant at each site. Choose the strongest plant and pull out the others. Tie the plant to the stake to give it good support. Trim off any side shoots, up to 60–90 cm from the ground.

Weeding

The area around the tree ($1\frac{1}{2}$ times the area of the canopy) should be kept clear of weeds for the first two years. After this, grass should be allowed to grow, but it should be kept short so that the fruits can be seen when they fall.

Any vegetation between the trees should be controlled. This can be done by grazing with animals when the trees are 3–4 years old, but not before then as the animals may break down the young trees.

Harvesting

Cashew trees normally begin to bear fruit when they are $2\frac{1}{2}$–3 years old. The trees reach maturity when they are 9–10 years old and they may have an economic life of 30–40 years if well cared for.

Only harvest nuts that have fallen from the tree. Remove the apple by a twisting action and cut away any of the apple that remains attached to the nut. Dry the nuts in the sun for a few days. In the dry season harvesting should be done once a week, but when it is wet you should harvest the nuts every day to avoid them rotting or being damaged by insects.

Only store the nuts when they are completely dry. Store them in clean, dry sacks in a well-ventilated place.

With good management, yields of 800–900 kg per hectare can be obtained, but the average yield is about 600 kg per hectare in pure stands. On many farms, yields are less because the cashew is intercropped with coconuts, mangoes, bananas and citrus.

PESTS AND DISEASES	SYMPTOMS	CONTROL
Cashew Helopeltis *Helopeltis* *anacardii*	These bugs suck the young leaves, fruit and buds. The leaves and buds turn brown and eventually die.	Pick off the bugs by hand, or spray with Gamma BHC or Parathion methyl. Before spraying, harvest as many of the nuts as possible.

Figure 86

Actual size

Coconut Bug *Pseudotherapterus* *wayi*	As for Cashew Helopeltis.	see above.

Citrus
Citrus spp

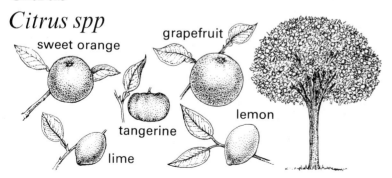

sweet orange

grapefruit

tangerine

lemon

lime

Introduction

Citrus fruits include lemons (*Citrus limon*), limes (*Citrus aurantifolia*), grapefruit (*Citrus paradisi*), sweet oranges (*Citrus sinenesis*), sour or Seville oranges (*Citrus aurantium*), mandarins, satsumas and tangerines (*Citrus reticulata*), and pummelos (*Citrus grandis*). Commercial production of citrus fruit is concentrated in sub-tropical countries with a 'Mediterranean' climate. The tropical countries produce only a small crop, mainly for home use.

As well as being eaten fresh, citrus fruit are used to make juice and marmalade, and oils are extracted from the skin for use in perfumes and confectionery.

World production

Oranges

Brazil	9,882 thousand tonnes per year
India	1,100
Mexico	3,240
Egypt	868
Argentina	697
Morocco	606
Total world production	37,883

Climatic range

Citrus trees require at least 900 mm of rain per year, and this should be evenly distributed over the growing season. Where this does not occur, irrigation is necessary to obtain high yields of good quality fruit.

Citrus fruits will tolerate high temperatures as long as there is enough water available, but they cannot withstand very low temperatures. The growth of citrus trees is greatly reduced if the temperature falls below 12°C (54°F) at any time of the year, and damage will occur if the temperature falls below 3°C (37°F).

However, fairly cool nights are necessary for the development of good colour in citrus fruit, especially sweet oranges. They will remain green, even when ripe, if grown in areas where the nights are warm. Upland areas therefore provide the best growing conditions for citrus fruit in the tropics. Limes, grapefruit and pummelos are better suited to lowland tropical areas than other types of citrus fruit.

Soil

Citrus trees can be grown on a wide range of soil types. However, a deep, light, loamy soil of good fertility is ideal. They do best in soils with a pH between 5·0 and 7·0.

Citrus trees cannot tolerate water-logged or saline (salty) soils.

Varieties

There are many good varieties and below are listed a few as examples. Your local research station should be able to advise you.

	Grows best at altitudes of
Sweet orange	
Washington Navel	1,000–1,800 metres
Valencia	0–1,500 metres
Hamlin	0–1,500 metres
Pineapple	0–1,500 metres
Grapefruit	
Marsh Seedless	0–1,500 metres
Red Blush	0–1,500 metres
Duncan	0–1,500 metres
Thomson	1,000–1,500 metres
Lime	
Mexican	0–1,500 metres
Tahiti	0–1,500 metres
Bearss	0–1,500 metres
Lemon	
Eureka	1,000–1,500 metres
Lisbon	1,000–1,500 metres
Villafranca	1,000–1,500 metres
Rough Lemon	0–1,800 metres
Mandarin	
Satsuma	0–1,500 metres
Kara	0–1,500 metres
Tangerine	
Clementine	0–1,500 metres
Dancy	0–1,500 metres

Planting

Commercial citrus trees are usually grown from cuttings, grafted on to a rootstock, except for tangerines which are often grown from seed. (Other citrus fruits will grow from seed, but the trees that are

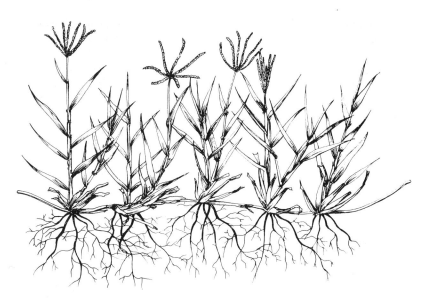

Figure 87   Couch grass

was made remains well above the ground. Lay a plank of wood on the ground, half-covering the planting hole, to show where ground level is: this will help you to set the tree at the correct height. Fill the hole with the manure/topsoil mixture, making the soil firm around the roots.

After planting, the subsoil can be used to make an irrigation ring around the tree, as shown in Figure 1 on p. 2. If planting out is done during the dry season, the tree should be watered well, and mulched to reduce water loss. About 35 litres of water per tree per day is needed. Provide a stake for support and tie the young tree to it.

Intercropping

For the first few years, low annual crops can be grown between the young citrus trees. This should only be done if there is an adequate supply of water and nutrients.

Wind protection

Citrus trees are damaged by strong, dry winds, which may cause the top branches to die back. A windbreak should be provided in areas where there are occasional high winds, or strong prevailing winds.

Irrigation

Citrus trees need moist soil throughout the year, but their water requirement reaches a peak at the time between flowering and ripening. Lack of water during this period will cause the flowers and fruit to fall off, so irrigation is essential in dry areas. Citrus fruit are a valuable crop and buying irrigation equipment will usually be worthwhile.

Weeding

The reasons for cultivation are to control weeds which compete with trees for fertility and moisture, and to incorporate cover crops and bulky organic materials into the soil. However, cultivation destroys surface citrus roots, affects soil structure unfavourably, and caused water to penetrate less readily.

Because of the disadvantages of tillage and the advantages of non-tillage, some citrus orchardists now control weed growth with weedkillers.

produced take a long time to bear fruit, and they may not show the characteristics expected of their variety.) Rootstock are chosen for their vigour and ability to withstand soil-borne diseases. The following rootstocks are often used: Rough Lemon, Sweet Orange, Cleopatra Mandarin and Troyer Citrange. Most growers obtain their trees already grafted from commercial nurseries.

Most citrus trees are planted at a spacing of 6 metres × 6 metres. The larger varieties, such as Valencia oranges and Lisbon lemons, are spaced at 8 metres × 8 metres, while grapefruit are spaced at 9 metres × 9 metres.

Clear all bushes and tree stumps from the area to be planted and remove perennial weeds such as couch grass (see Figure 87) which may become troublesome later. Prepare the planting holes well in advance of planting, before the rainy season.

Dig holes 60 cm wide and 60 cm deep, keeping the topsoil and subsoil separate. Mix the topsoil from each hole with $1\frac{1}{2}$ 20-litre drumfuls of well-decomposed manure or compost and 250 gm of double superphosphate, (1 kg tinful for every 4 trees).

Plant the young trees slightly higher than they were in the nursery to allow for settling, and to ensure that the point where the graft

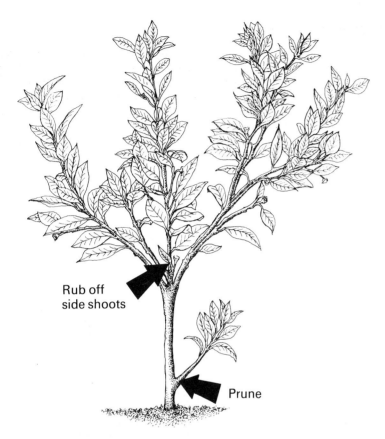

Figure 88   Pruning a citrus tree

Pruning

All side shoots growing from the rootstock and from the tree trunk below the main branches should be rubbed off. This should be done immediately after planting and repeated at regular intervals, (see Figure 88). Additional pruning of the young tree is not usually necessary, since citrus trees normally assume a satisfactory shape of their own accord.

The height at which branching occurs should be about 1 metre above ground level. If branching does not take place when the tree has reached this height, cut off the top of the stem.

Remove dead and broken branches. Prop up branches that are sagging with the weight of the fruit.

Manure and fertilizer

Manure or compost should be applied, at a rate of one 20-litre drumful per tree per year.

Nitrogen fertilizer should be applied every year, and phosphates from the third year onwards. The amounts are shown in the table below. Apply the fertilizer at the beginning of the rainy period. If there are two rainy periods in the year, divide the quantity into two, and apply half at the beginning of each rainy period.

Potash is important as a fruit sweetener. Applications of up to 750 gm of sulphate of potash per year for a mature tree, in split applications, will improve fruit quality if there is a deficiency. Soil testing will show whether an application is necessary.

	Double superphosphate, per tree	*CAN, per tree*
Year 1	none	100 gm (5 large spoonfuls per tree)
Year 2	none	250 gm
Year 3	125 gm per tree (6 large spoonfuls per tree, or 1 kg tinful for every 8 trees)	400 gm
Year 4	250 gm per tree (1 kg tinful for every 4 trees)	550 gm
Year 5	250 gm per tree (1 kg tinful for every 4 trees)	700 gm
Year 6	500 gm per tree (1 kg tinful for every 2 trees)	1 kg
Year 7 onwards	500 gm per tree (1 kg tinful for every 2 trees)	1–2 kg

Nutrient deficiency

One or more nutrients may be deficient in citrus orchards. The usual symptom is leaf discoloration, but the symptoms vary depending on which nutrient is deficient. Leaves with typical deficiency symptoms are shown in Figure 89, and a detailed list of symptoms is given overleaf. This may enable you to identify which nutrient is deficient in your orchard. However, the best way to check which one is deficient is to have a leaf sample tested at a testing station or laboratory.

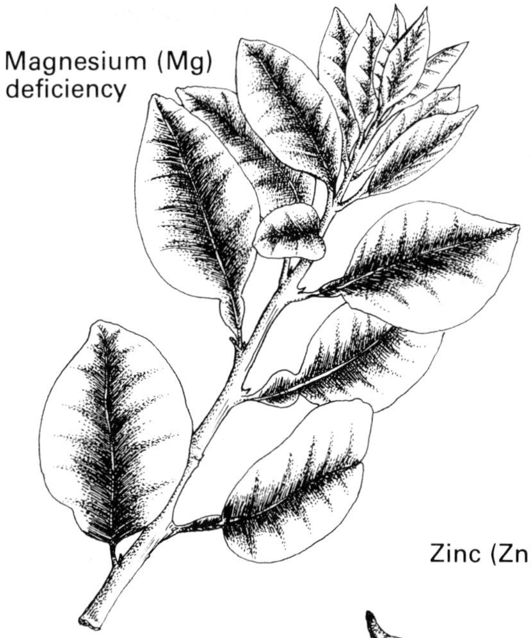

Magnesium (Mg) deficiency

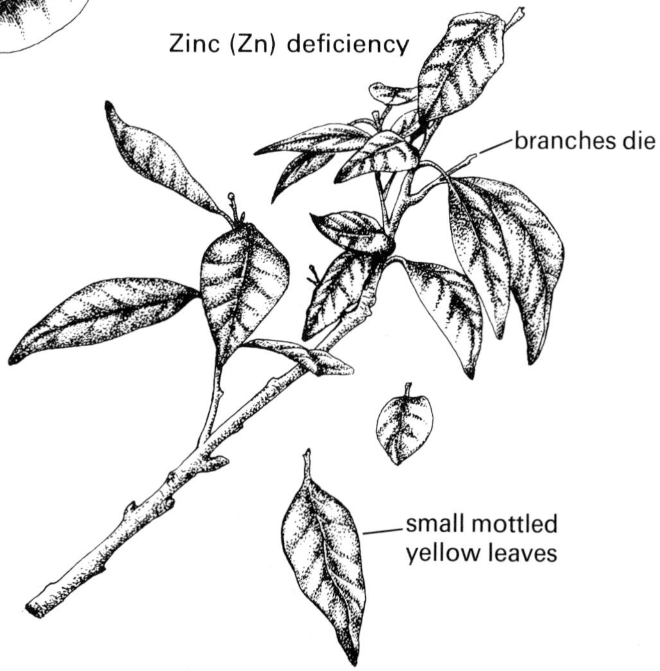

Copper (Cu) deficiency

Zinc (Zn) deficiency

branches die

large dark leaves

small mottled yellow leaves

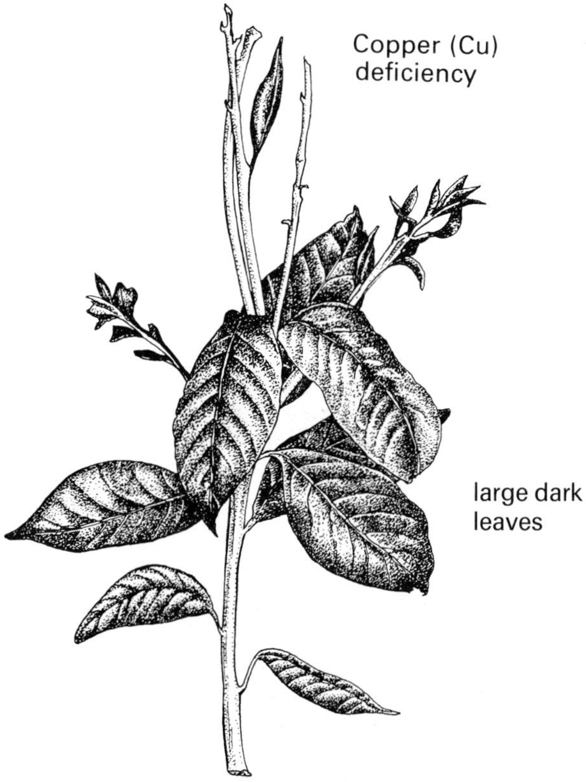

Figure 89   Symptoms of mineral deficiencies in citrus trees

Nutrient	Symptoms	Treatment	Nutrient	Symptoms	Treatment
Nitrogen (N)	Leaves become pale yellow to yellowish-white in colour, with the small veins also yellow. Tree growth is reduced, and although many flower buds are produced, these fall without opening, so that the yield is low.	Apply nitrogen fertilizer (CAN) using the amounts shown.		become brownish-yellow or metallic in colour. Gum pockets appear on both the fruit and leaves. Many of the young fruit fall off. Tree growth is reduced and the tree has a 'cabbage-head' shape.	
Phosphates (P)	The leaves are small and dull. Tree growth is reduced and the yield is low. The fruit are large, with a puffy, bumpy surface, thick rind and an enlarged central cavity.	Apply double super-phosphate using the amounts shown.	Copper (Cu)	The leaves become very large, and turn dark green in colour. The fruit splits, or has dark, gum-soaked eruptions and gum in the central core. The fruit may also turn black. Many short, weak twigs are produced. The twigs enlarge at the nodes, then blister and die. There are gum pockets on the twigs and branches. Tree growth is reduced and the tree has a 'cabbage-head' shape.	Apply a top-dressing of copper sulphate, using 30–100 kg/ha. Alternatively, use copper fungicides instead of other fungicides, to control fungal diseases.
Potash (K)	The old leaves curl and lose their green colour. Tree growth is reduced and the fruit drop before they should. They are small, with a thin smooth rind, and not very sweet.	Apply sulphate of potash, up to 750 gm per tree per year, as a top-dressing.			
Magnesium (Mg)	The leaves become mottled yellow along the margin, so that only a wedge-shaped segment in the centre is green. Eventually they become completely yellow and fall off. The tree growth and yield are both reduced.	Apply magnesium sulphate to the soil, using about 100 ka/ha. This can be added to other fertilizers.	Iron (Fe)	The leaves are small and turn yellow between the veins; eventually the veins turn yellow too. Tree growth and yield are reduced.	Apply a top-dressing of iron sulphate, using about 30 kg/ha.
Boron (B)	The leaves curl and pucker and develop corky veins. They	Apply borax as a top-dressing, using about 10 kg/ha.	Manganese (Mn)	The outer parts of the leaf turn pale green or light yellow, except around the veins. The yield is reduced and	Spray the leaves with manganous sulphate, using about 10 kg/ha.

Nutrient	Symptoms	Treatment
	eventually the tree growth is also reduced.	
Zinc (Zn)	The leaves become mottled yellow between the veins. They are small and fall off earlier than usual. Twigs and small branches die. The yield is reduced and some of the fruit are pale, small and thick-skinned. Tree growth is eventually reduced.	Spray the leaves with zinc sulphate, using about 5 kg/ha.

Commercial liquid fertilizers, often called 'foliar feed', can also be used to correct deficiencies of minor nutrients. These are sprayed onto the leaves.

Harvesting

The first economic crop can be expected 3–4 years after transplanting. There are usually two citrus-picking seasons each year. Yields vary according to the variety, but the following yield ranges can be expected:
For oranges: 90–130 kg per tree
For lemons: 130–180 kg per tree

Citrus fruit should be picked when they are fully mature and the same colour all over. In areas where the fruit remains green, a few should be picked to see if they are ripe.

Pick the fruit one by one, by clipping off the stalk or by bending and pulling it with a slight twist. Take care not to injure or bruise the fruit, as this will lead to rapid spoilage.

Pests and diseases

Citrus trees suffer from many diseases and are attacked by a great many pests. You should inspect the orchard carefully, at least twice each week, for any signs of pests or diseases. Fallen fruit should be collected and buried to prevent it from becoming a source of disease and insect pests.

PESTS AND DISEASES	SYMPTOMS	CONTROL
Citrus Aphid *Toxoptera citricidus* and *T. aurantii*	Clusters of small black or brown soft-bodied insects found under the leaves and on young shoots. They suck plant sap and cause distortion of the leaves. Honeydew (a sticky liquid produced by the aphids) and sooty mould is often present. Aphids can carry Tristeza Disease from one plant to another.	Spray with Diazinon, Malathion, Fenitrothion or Dimethoate.

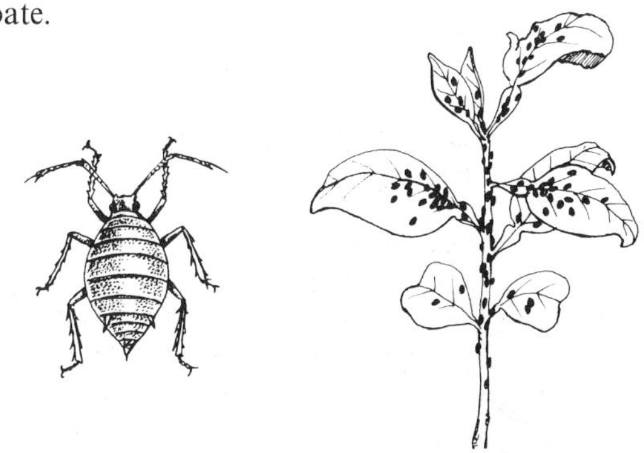

Figure 90

Actual size

113

PESTS AND DISEASES	SYMPTOMS	CONTROL

Citrus Psyllid
Trioza erytreae

Small, brown, scale-like insects which settle underneath the leaves, causing pits on the underside and bumps on the upperside. The leaves are cupped or otherwise distorted, and yellowish in colour.

As for Citrus Aphids.

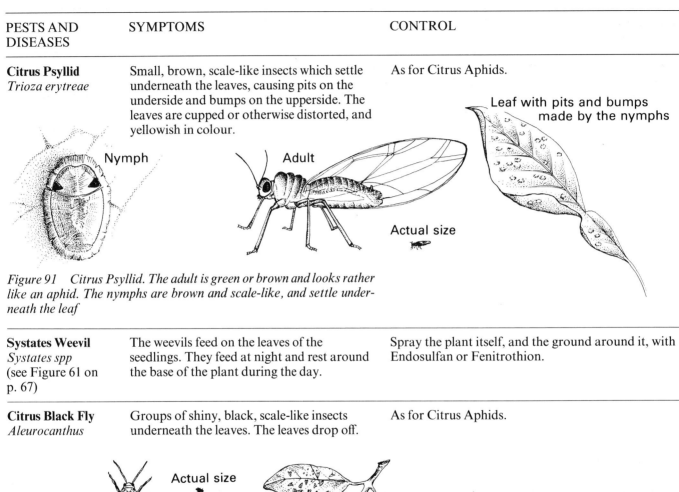

Figure 91   Citrus Psyllid. The adult is green or brown and looks rather like an aphid. The nymphs are brown and scale-like, and settle underneath the leaf

Systates Weevil
Systates spp
(see Figure 61 on p. 67)

The weevils feed on the leaves of the seedlings. They feed at night and rest around the base of the plant during the day.

Spray the plant itself, and the ground around it, with Endosulfan or Fenitrothion.

Citrus Black Fly
Aleurocanthus

Groups of shiny, black, scale-like insects underneath the leaves. The leaves drop off.

As for Citrus Aphids.

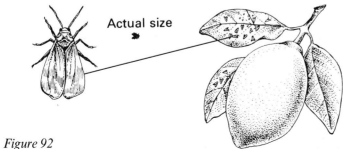

Figure 92

PESTS AND DISEASES	SYMPTOMS	CONTROL
Scale Insects **Red Scale** (see Figure 83)	Small, flat, scale-like insects with no visible legs, which attach themselves firmly to the leaves, branches and fruit. They are often associated with honeydew (a sticky liquid) and black, sooty mould; ants may be present.	Harvest as much fruit as possible, then spray with a mixture of Fenitrothion and White Oil, Malathion and White Oil, or Diazinon and White Oil. Repeat the spray after 2–3 weeks if necessary.

Figure 93   Scale Insects

Mussel Scale (enlarged)

Red Scale (enlarged)

| **False Codling Moth**
Cryptophebia
leucotreta | The moth lays its eggs on the fruit, and when the caterpillars hatch out they bore into the pulp. A yellow patch can be seen on the skin of the fruit where the caterpillar has entered. | Collect all infested fruit from the ground, pick all infested fruit from the tree, and bury it. It must be buried at least 50 cm deep.
As a preventive measure: Where this pest is prevalent, spray with Fenthion or Malathion once a week, starting when the fruit measure about 4 cm across. |

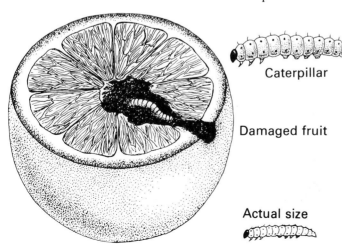

Caterpillar

Damaged fruit

Figure 94

Actual size

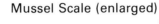

PESTS AND DISEASES	SYMPTOMS	CONTROL

Orange Dog
(also called the
Citrus Swallowtail
or Christmas
Butterfly) *Papilio
demodocus*

This butterfly lays its eggs on the leaves of the plant, and the caterpillars eat the leaves, often causing serious damage.

If there are not too many caterpillars, remove them and the pupae by hand and kill them.
If the trees are heavily infested, spray with Fenthion, Fenitrothion, Diazinon or Malathion.

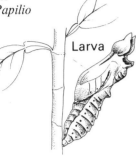

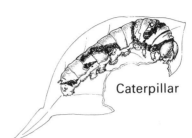

Figure 95   Orange Dog (actual size). The butterfly is dark brown with many pale yellow spots, and two orange spots on the hindwings. The caterpillar is pale green with brown markings. The pupa is green or brown

Fruit Flies
Ceratitis capitata
and *Ptendarus rosa*

The flies lay their eggs on the fruit and when the larvae hatch out they tunnel into the fruit, turning the flesh brown and rotten. Small dark spots can be seen on the skin of the fruit. If unripe fruit is infested it may fall off before it is ready.

Bury all infested fruit.
As a preventive measure: If the pest is prevalent, spray routinely with Fenthion or Fenitrothion, mixed with protein hydrolysate, starting as soon as the fruit sets. There is no need to spray every row of trees: spraying alternate rows will be sufficient to control the pest.

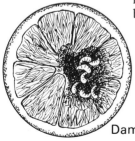

Actual size

Figure 96

Citrus Bud Mite
Aceria sheldoni

These tiny pests attack the growing shoots. The twigs, leaves and flower buds become deformed. The fruit may also be affected.

As soon as any signs of attack are seen, spray with Dicofol, Chinomethionate, Binapacryl, Dimethoate, or Sulphur.

PESTS AND DISEASES	SYMPTOMS	CONTROL
Citrus Rust Mite *Phyllocoptruta oleivora*	These tiny pests attack the fruit. Lemons become silvery in colour, while oranges and grapefruit become russet-coloured. The skins of the fruit are thicker than usual.	As for Citrus Bud Mite. (Harvest as much of the fruit as possible before spraying.)
Red Spider Mite *Tetranychus spp*	These tiny pests attack the leaves, which become mottled with silver and brown. White droppings can be seen, especially underneath the leaves. If the trees are severely infested, the leaves may drop off.	As for Citrus Rust Mite.
Fire Ant *Solenopsis geminata*	See p. 104	
Leaf-footed Plant Bug *Leptoglossus australis*	See p. 46	
Tristeza Disease	This is a viral disease that attacks citrus trees growing on non-resistant rootstocks. The tree does not grow well, the leaves become bronzed in colour and fall off and the twigs die. When a small strip of bark is removed just below the graft union, a honeycomb-like pattern can be seen. Eventually the tree dies.	There is no chemical control for viral diseases. The disease is carried from one tree to another by Aphids, so spray these pests to prevent the disease spreading. *As a preventive measure:* Use resistant rootstocks such as Rough Lemon, Sweet Orange, or Cleopatra Mandarin.
Citrus Scab *Elsinoe fawcetti*	Yellow–orange corky spots appear underneath the leaves, singly or in groups. The leaves may become crinkled or rolled. Similar spots also appear on the twigs and fruit.	Burn all infected trees. *As a preventive measure:* If the disease is prevalent, spray once every two weeks with a copper fungicide. Begin when the new leaves appear and continue until flowering has ended.
Green Mould *Penicillium digitatum*	This disease attacks stored fruit, producing a green–grey mouldy growth on the skin of the fruit. It may also attack the crop in the orchard, mainly fallen fruit.	When harvesting, take care not to bruise the fruit as this encourages the growth of the mould. Store the crop in a well-ventilated place. Inspect it regularly and throw out any infected fruit.

PESTS AND DISEASES	SYMPTOMS	CONTROL
Citrus Greening Disease	The leaves are small and yellow, or mottled yellow and green. (The mottling may look similar to that caused by zinc deficiency, see p. 111.) The leaves tend to point upwards at an unusual angle. This is very serious disease, which is carried from one tree to another by Citrus Psyllids.	There is no chemical control for this disease, except using very expensive methods which are not worthwhile for the small grower. *As preventive measures:* Raise seedlings in a nursery area well away from the orchard. Spray the seedlings to kill Citrus Psyllids, as described on p. 114.
Leaf Spot *Alternaria citrii*	This disease causes large blotches with a yellow halo on the leaves. It also attacks the stored fruit, causing it to rot. The disease is more common at high altitudes.	If leaf spots develop, or if the disease is prevalent, spray once every two weeks with a copper fungicide, starting when the new leaves appear. *To control the disease in stored fruit:* Store the crop in a well-ventilated place. Inspect it regularly and throw out any infected fruit.

Coconuts
Cocos nutrifera

Introduction

Coconut palms are an important crop in many coastal areas of the tropics. The mature nuts contain a thick white flesh, which is dried and sold as copra. Oil can be extracted from copra for the manufacture of cooking oil, margarine and soap.

Coconuts are also sold immature, when the flesh is eaten and the juice (coconut milk) is drunk. Coconut palms are also tapped for palm wine: the unopened flower is tapped for the sap which, after a few days, turns into an alcoholic drink. Palm wine production can be quite profitable, but over-production weakens the trees. The trunks of coconut palms are used for house- and boat-building, and for firewood. The dried leaves are used for thatching, and they can also be made into mats or fences. The coarse fibre found on the outside of the nuts is called coir, and it is used for matting and upholstery.

World production

Indonesia	10,800 thousand tonnes per year
Philippines	8,918
India	4,300
Sri Lanka	1,700
Mexico	788
Mozambique	400
Total world production	33,968

Climatic range

Coconuts do well in hot areas, and require a rainfall of at least 1,250 mm per year. They therefore grow best at altitudes below 1,000 metres.

Soil

Coconuts need a well-drained soil. They do well on rather poor, sandy soils, and will tolerate a certain amount of salt in the soil, but they do not grow well on shallow soils over coral.

Varieties

The tall variety is the one most usually planted, as it gives the best yield. Dwarf varieties are easier to harvest but do not yield as much.

Planting

Obtain nuts for planting from healthy trees. Prepare trenches 15 cm deep and 50 cm apart. Bury the nuts in these trenches, about 25 cm apart. Water in well, and keep the soil moist until the seedlings are established.

Figure 97   Only the strong seedlings on the left should be planted. Good seedlings have strong, straight stems and large leaves. Bad seedlings have small, twisted stems and small leaves

When the seedlings have three or four leaves (after about one year) they are ready for transplanting. Select only the most vigorous and healthy seedlings and throw away the others, (see Figure 97). About half the seedlings should be thrown away.

Transplant the seedlings at the beginning of the rainy season. The spacing should be 9 metres × 9 metres.

Dig holes 60 cm wide and 60 cm deep, keeping the topsoil and subsoil separate. Mix the topsoil from each hole with a 20-litre drumful of manure or compost, and use this mixture to refill the hole. The nut should be placed 30 cm below the surface of the soil, and the stem earthed up as it grows.

Take great care not to damage the roots when transplanting.

Intercropping

Some growers adopt a wider spacing and interplant the coconuts with cashew, mangoes, bananas or cassava. This may have some advantages in helping to control pests. If the coconuts are grown alone and the spacing recommended here is used, then intercropping with annual crops can be done for the first two years without any damage to the coconuts.

Fertilizer

Apply CAN, at a rate of 1 kg per tree per year, beginning in the first year after transplanting. Spread the fertilizer in a ring around the base of the tree.

Weeding

Coconuts will not grow well if there is too much competition from weeds. However, weeding with mechanical implements may damage the roots and lead to disease. Controlling the weeds with fire can also damage the trees. The weeds should therefore be cut down by hand, or kept down by cattle.

Replacement of trees

The economic life of a coconut plantation is about 60 years, but to maintain high yields, any dead, diseased or low-yielding trees should be removed. Replace these trees with new seedlings.

Harvesting

Nuts for copra need to be mature and they are ready when they fall to the ground. If selling fresh nuts, these should be harvested at intervals of two months. They are ready to pick about eleven months after flowering; an experienced grower will be able to tell exactly when they are ready.

Under ideal conditions, and with high rainfall, each tree may yield over 100 nuts per year.

Drying copra

Copra can be produced either by sun-drying or kiln-drying. Sun-drying is done as follows:

The nuts are cut in half and drained of water, then spread out in the sun, cut side uppermost. After 2–3 days the flesh is removed and this is then dried in the sun for another 4–5 days. At night, and during rain, it is covered up. Continuous sunshine is needed for the production of high-quality copra.

PESTS AND DISEASES	SYMPTOMS	CONTROL
Rhinoceros Beetle *Oryctes monoceros*	The beetles attack the buds in the crown of the coconut palm, producing V-shaped holes in the palm fronds. If the growing point is eaten the tree dies.	There is no worthwhile chemical control for this pest. The beetle lays its eggs in decaying matter, especially rotting coconut trunks, so the best way of controlling it is to keep the plantation clear of all refuse. Split and burn all old coconut trunks. This is very important.

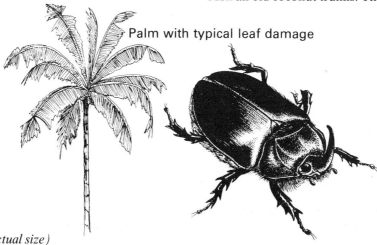

Palm with typical leaf damage

Figure 98   Rhinoceros Beetle (actual size)

PESTS AND DISEASES	SYMPTOMS	CONTROL

Coconut Bug
Pseudotheraptus wayi

Actual size

The bug punctures the young nuts and they drop off.

Nymph

Adult

There is no effective chemical control for this pest. Some ants eat this bug, and they can be encouraged by interplanting citrus or cashew trees with coconut palms. These ants are preyed on by other ants which reach them by walking up the trunks of the trees. Spray a band of Dieldrin or Gamma BHC around the trunk of each tree. This will keep these ants off the tree, and allow the other ants to increase in number and thus eat more of the Coconut Bugs.

Figure 99   Coconut Bug. The bug is brown in colour with red markings

---

Coconut Scale
Aspidiotus destructor

Yellowish scales (see Figure 93 on p. 114) found underneath leaves, on flowers and on young nuts. If severely infested, the leaves turn yellow and die.

Spray with Malathion, Parathion methyl or Diazinon.
As preventive measures:
1. Do not plant the palms too close together.
2. Practise field hygiene.

Coconut Palm Weevils

Rhynchophorus spp

Dig up and burn all infested palms. Spray the trunks of the remaining palms with Aldrin or Dieldrin.
As preventive measures:
1. Avoid injuries to the palm trunks.
2. Practise field hygiene.

The weevils bore into the trunk of the coconut palm.

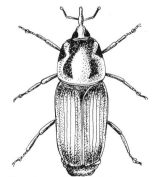

R. phoenicis Africa
reddish-brown

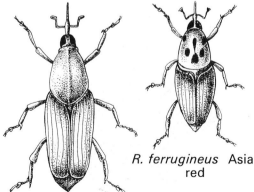

R. palmarum South America
brown-black

R. ferrugineus Asia
red

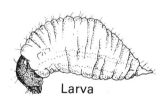

Larva

Figure 100   Coconut Palm Weevils (actual size)

PESTS AND DISEASES	SYMPTOMS	CONTROL
Bole Rot *Marasmiellus cocophilus*	This is a fungal disease that is carried in the soil and enters the plant through the roots. It causes the stem to rot, and the fronds to turn yellow and wilt. This disease attacks seedlings and young palms with damaged roots. The plants eventually die.	There is no chemical control for this disease. Dig up and burn all infected plants. These plants can be replaced with new seedlings, but transplant them very carefully, to avoid damaging the roots. Weeding should also be carried out with great care.
Bud Rot *Phytophthora palmivora*	The soft growing point of the palm becomes rotten. The young fronds wither and the palm eventually dies.	Cut down all infected trees immediately and burn them. They can be replaced with new seedlings.
Leaf Spot *Helminthosporium spp*	This disease attacks the seedlings, producing spotting on the leaves.	Spray the seedlings with a fungicide, such as Benomyl.
Cadang-cadang	The coconut palms become less and less vigorous, and produce smaller and smaller fronds. The cause of this disease is unknown.	There is no known cure for this disease. Some palms seem to be resistant, and seed should be taken from such palms for replanting.

Guava
Psidium guajava

Introduction

Guavas are eaten fresh, or they may be canned, or made into jam or juice.

Guavas are the fruit of a tree which grows to a height of 3–10 metres. Another name for the guava is guayaba.

Climatic range

The guava tree can be grown in a wide variety of climates. It is drought resistant and also tolerates flooding. It can withstand high temperatures, but does not do well in low temperatures and is damaged by frost. For commercial production, therefore, areas below 1,000 metres are the most suitable, although the tree can be grown up to 1,500 metres. For good yields, the rainfall should be between 1,000 mm and 2,000 mm per year.

Soil

The guava tree tolerates many different soils, and can even be grown in infertile, acid or water-logged soil.

Varieties

Improved varieties are now available, that have larger fruit with more pulp, thin skins and fewer seeds. Some improved varieties are entirely seedless, while others have white flesh instead of red.

Improved varieties must be vegetatively propagated (grown from grafts) to preserve the characteristics of the variety.

Growing from seed

Guavas are often grown from seed, but this is not recommended as the characteristics of the trees are very variable when grown from seed. The fruit produced varies a great deal in size, shape, sweetness, flavour and colour. However, the rootstocks for grafting are obtained by growing from seed.

If guavas are grown from seed, the seed should be dressed with a fungicide such as Thiram or Captan before sowing. Sow the seed 1 cm deep in nursery beds with a light soil and later transplant into pots or plastic bags. Alternatively, the seed can be sown directly into pots. The soil must be kept moist. After 5–7 months the seedlings should be ready for transplanting, but if necessary they can be kept in pots for up to one year.

Grafting

Bud grafting is the best way of producing new trees, since it gives more uniform trees, with known characteristics.

Seedlings grown from seed are used for the rootstock. A healthy and high-yielding tree is chosen to supply the budwood, and the leaves and branches removed about twelve days before grafting, to encourage the buds to develop. Expert advice should be sought on how to make the graft.

Transplanting

The young trees should be planted out at a spacing of 7 metres × 7 metres. Manure or compost should be worked into the soil before transplanting if possible, and double superphosphate added to each planting hole at a rate of 60 gm (3 large spoonfuls) per hole.

Fertilizer

Apply double superphosphate when transplanting as described above.

Top dress with CAN, applying 60 gm (3 large spoonfuls) per tree in the first year, and 120 gm (6 large spoonfuls) per tree in the following years. If the trees are very productive, more fertilizer can be applied, up to about 300 gm per tree (15 large spoonfuls per tree or just under 1 kg tinful for every 3 trees). Split the fertilizer into 3 equal amounts and apply one portion every 4 months.

Pruning

Pruning can be done to maintain the shape of the trees and to remove dead wood. Suckers growing from the rootstock should be removed.

Harvesting

The trees first bear fruit about two years after transplanting, and the yields increase for several years after this. The economic life of the trees is about thirty years. The fruit should generally be picked when ripe, but for export it can be picked just before it is ripe.

Mature trees yield about 12–16 tonnes per hectare per year, while improved varieties can yield as much as 40–50 tonnes per hectare per year.

PESTS AND DISEASES	SYMPTOMS	CONTROL
Fruit Flies *Dacus dorsalis, Ceratitis capitata* or *Anastrepha spp* (see Figure 96)	The flies lay their eggs on the surface of the fruit, and when the larvae hatch out they bore into the fruit.	If fruit flies are prevalent, spray routinely with Malathion, Fenthion or Trichlorphon.

PESTS AND DISEASES	SYMPTOMS	CONTROL
Striped Mealybug *Ferrisia virgata* and **Kenya Mealybug** *Planococcus kenyae* (see Figure 60 on p. 66)	Masses of soft-bodied, pale insects. They feed on the young shoots, fruit and leaves, weakening the trees and reducing the yield.	*If the pest is present:* Spray the leaves with Diazinon, Malathion or Dimethoate. *As a preventive measure:* Spray a band of Dieldrin or Gamma BHC around the trunk of each tree. The band should be 15 cm wide. This will keep ants out of the tree. Normally the ants tend the Mealybugs and protect them from other predators; without the ants there, the predators can eat the Mealybugs.
Scale Insects: Soft Green Scale *Coccus alpinus* and *C. viridis* **Helmet Scale** *Sansettia coffeae* **Cottony Cushion Scale** *Icerya purchasi*	Flattish, scale-like insects with no visible legs which cling to the leaf. (See Figure 93 on p. 115). They are found in rows along the veins, underneath the leaves. On the upperside of the leaves, spots of honeydew (a sticky liquid) are found.	As for Mealybugs.
Red-banded Thrips *Solenothrips rubrocinctus* (see Figure 102 on p. 127)	The leaves become dark-stained and rusty in appearance. Small black droppings may be seen.	Spray with Fenthion or Fenitrothion.
Root Rot *Clitocybe tabescens*	The roots and crown of the tree go rotten.	Dig up and burn all infected trees.
Anthracnose *Colletotrichum gloeosporioides*	Mature fruits go brown and rotten.	If the tree is severely infected it should be dug up and burned. If infection is only slight, spray with Benomyl. *As a preventive measure:* Apply more nitrogen fertilizer.

PESTS AND DISEASES	SYMPTOMS	CONTROL
Leaf Blotch *Glomerella cingulata*	This disease attacks the leaves and young fruits, which go black and rotten.	Spray with Benomyl.

Mangoes
Mangifera indica

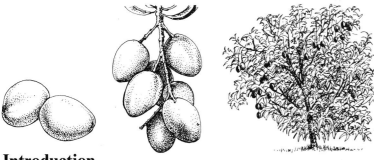

Introduction

The mango is a large evergreen tree, some varieties of which may reach a height of 17 metres, and can live for over 100 years.

Mangoes are eaten fresh or canned, and are also used to make fruit drinks. Green mangoes are used for making pickles and preserves.

World production

India	9,300 thousand tonnes per year
Brazil	680
Pakistan	600
Mexico	570
Philippines	338
Tanzania	172
Zaire	172
Total world production	14,034

Climatic range

Areas with a hot summer and a cooler winter are ideal for mangoes. There should be little variation between day and night temperatures. Mangoes therefore grow best at altitudes below 1,500 metres.

Mangoes can grow when the rainfall is as low as 650 mm per year, but they do better on a higher rainfall, about 1,500 mm per year being ideal.

Soil

The soil should be well drained and deep but does not need to be very fertile.

Varieties

There are many varieties grown in different parts of the world. Below are listed some examples.

Ngowe — This variety has a long, deep yellow fruit of good quality, and is a popular variety for export.

Boribo — This variety has large, long, orange-red fruit of good quality that are especially suitable for canning. A high-yielding variety.

Batawi — The fruit of this variety are very large and round, ranging from olive green to purple in colour, and of good quality.

Apple — A good variety for export. The fruit are round and yellow-orange to red in colour.

Dodo — A high-yielding variety, but the fruit are rather fibrous.

The following varieties can withstand cooler conditions, and are therefore suitable for growing at altitudes up to 1,800 metres: Sabre, Harries and Peach.

Grafting

Ungrafted mango trees grow very large and so are more difficult to manage than grafted trees. The fruit is usually of an inferior quality. Grafted mangoes can be obtained from commercial nurseries.

Transplanting

Transplanting should be done at the beginning of the rainy season. Grafted trees should be spaced at 12 × 12 metres, and ungrafted trees at 14 × 14 metres. Dig planting holes 1 metre deep and 1 metre across, keeping the topsoil and subsoil separate. Mix the topsoil with a 20-litre drumful of manure or compost, and 60 gm (3 large spoonfuls) of double superphosphate. Use this mixture to refill the hole. Firm the soil around the plant, water in well and mulch.

The subsoil can be used to make an irrigation circle around the plant, as shown in Figure 1 on p. 2.

Irrigation

Irrigation is needed during the first year, but will probably not be necessary in subsequent years.

Pruning

In the first year, cut back the seedling to a height of 1 metre, to produce a spreading framework of branches.

In the second year, choose four or five well-spaced branches to become the main branches of the tree. Cut off all other branches.

After this, no more pruning is required, except to remove dead branches.

Fertilizer

Apply CAN fertilizer each year, using the quantities shown in the table below. In the first year, apply the fertilizer immediately after transplanting; in subsequent years apply it at the beginning of the rainy season.

Amount of CAN to apply per tree

Year 1	50 gm (2½ large spoonfuls)
Year 2	100 gm
Year 3	150 gm
Year 4 and all following years	200 gm

If potash is deficient in the soil, a compound fertilizer such as 12-15-18 can be applied instead of CAN. Apply 100 gm per tree in the first year, 150 gm per tree in the second year, 200 gm per tree in the third year and 300 gm per tree from the fourth year onwards.

Weeding

Clear excessive vegetation from beneath the trees, cutting it down by hand.

Harvesting

Mangoes begin to bear fruit four or five years after planting, and are in good production after eight years. They reach maturity at twenty years of age.

Each tree should produce about 200–500 fruit per year, and some varieties can produce up to 1,000 fruit each year. Most varieties are erratic in the amount they produce, and a poor harvest may follow a good one.

The fruit are ready to harvest when they change colour and when the first few ripe fruit drop from the tree. They should be harvested at regular intervals.

Handle the fruit gently when picking and packing. The fruit should be cut with 5–10 cm of stalk attached.

For export, cut the stalk to 2 cm and wash the fruit to remove any dirt and spray residues. Pack in boxes, in single layers, using plenty of woodwool or other packing material.

When harvesting, use ladders to avoid breaking the branches of the trees.

PESTS AND DISEASES	SYMPTOMS	CONTROL
Mango Seed Weevil *Sternochetus mangiferae*	The larvae of this weevil are fleshy white grubs up to 3 cm long. They enter the fruit and attack the seed. There may be a hard white area in the fruit; the fruit may fall early or rot in storage.	To prevent the spread of this pest keep the orchard clean and remove fallen fruit twice a week. 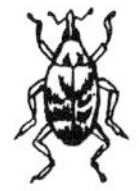 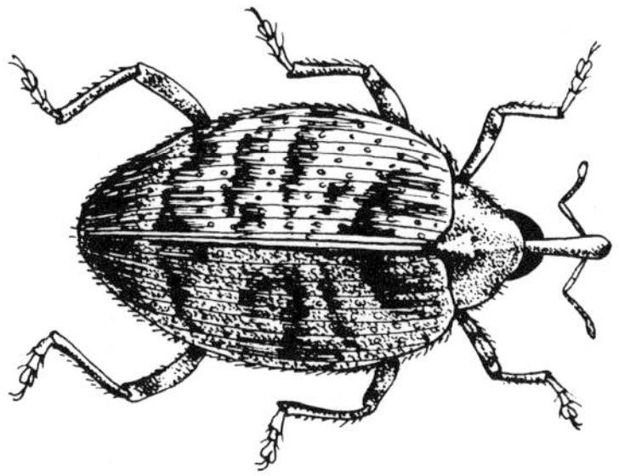

Figure 101   **Actual size**

Mango Scales (Many different species. See Figure 93 on p. 114 for typical Scale Insects.)	Small, flat, oval or circular scale insects found all over the tree and fruit. They produce honeydew (a sticky liquid) and are usually associated with a black, sooty mould.	Harvest as much fruit as possible and then spray with a mixture of Diazinon and White Oil, or Malathion and White Oil.
Fruit Flies *Ceratitis capitata* (see Figure 96 on p. 116).	The fly lays its eggs on the fruit, and shiny white maggots, up to 1 cm long, are found inside the fruit. The mangoes change colour before they are ripe and the flesh may become liquid.	Harvest as much fruit as possible; sort out the edible fruit and bury all those that are infested. Then spray with Fenthion or Fenitrothion, mixed with protein hydrolysate. Do not harvest again for two weeks. *As a preventive measure:* Collect fallen fruit regularly and bury it deeply in the soil.
Red-banded Thrips *Solenothrips rubrocinctus*	The leaves become dark-stained and rusty in appearance. Small black droppings may be seen.	Spray with Fenitrothion or Fenthion.

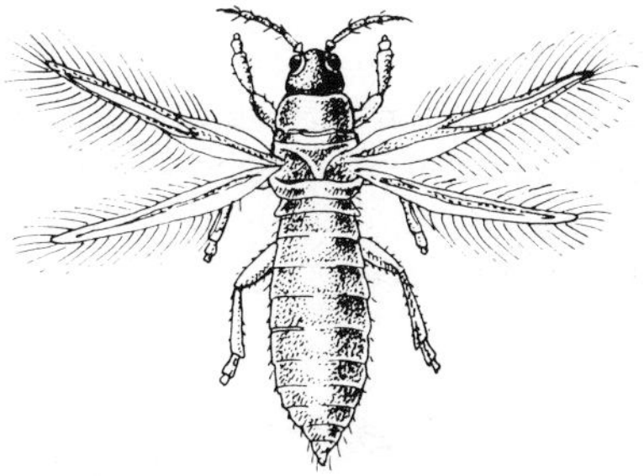

Figure 102

PESTS AND DISEASES	SYMPTOMS	CONTROL
Anthracnose *Colletotrichum gloeosporioides*	Black spots appear on the fruit.	Burn all infected fruit and other crop residues to prevent the disease from spreading. *As a preventive measure:* If the disease is prevalent, spray routinely with Captan or a copper fungicide, from just before flowering until harvest time.
Powdery Mildew *Oidium mangiferae*	A white mildewy growth appears, mainly on the flowers.	*If the plants are infected:* Spray with Dinocap. *As a preventive measure:* If the disease is prevalent, the trees can be sprayed with Dinocap once every 2 weeks, starting at flower-bud stage, and continuing until the fruit has set.

Rambutan
Nephelium lappaceum

Introduction

Rambutans are the fruit of a tree which grows to about 14 metres in height. They are also known as hairy lychees, since the fruit are similar to lychees. The pulasan (*Nephelium mutabile*) is closely related to the rambutan, and is cultivated in a similar way.

Climate and soil

Rambutan trees grow well in hot, humid areas, with over 2,000 mm of rain per year, and it is therefore best suited to lowland areas. The soil should be fertile and well drained.

Growing from seed

Growing from seed is not recommended as the trees produced may not have good characteristics, and they often bear sour fruit. They also take longer to bear fruit than grafted trees. However, seedlings to be used as rootstocks for grafted trees should be grown from seed.

Grafting

Seedlings about 5–6 months old should be used for the rootstock. Bud grafting is the best way of producing a uniform tree with good characteristics. Expert advice should be sought on how to make the graft. Trees may also be propagated by layering; seek expert advice on how to do this.

Planting

The spacing used depends on the type of tree being planted.
Bud-grafted trees need a spacing of about 12 metres × 12 metres. The young trees should be planted out when the soil is moist. Holes 60 cm deep and 60 cm wide should be dug. Keep the topsoil and subsoil separate and mix the topsoil with a 20-litre drumful of manure and 40 gm of double superphosphate (2 large spoonfuls). Use this mixture to refill the hole.

Fertilizer

Apply CAN every year, using the same quantities as recommended for mangoes (see the table on p. 126).

In the first year, apply the fertilizer immediately after transplanting and in subsequent years apply it at the beginning of the rainy season. It is particularly important to apply fertilizer if the tree is bearing a heavy crop of fruit.

If potash is deficient in the soil, a compound fertilizer can be applied instead of CAN. Seek local advice on how much to apply.

Cultivation

The ground should be kept clear of weeds, but be careful not to disturb the roots of the tree. Weed by hand or use a herbicide.

Pruning is not usually necessary, but the upper branches may be cut back so that the fruit are easier to harvest.

Harvesting

Grafted trees usually bear fruit after 2–3 years. There are often two harvests each year, so a mature tree can produce 5,000 fruit or more each year.

PESTS AND DISEASES	SYMPTOMS	CONTROL
Fruit Flies and **Mealybugs**	See pp. 123 and 124	
Blight *Phomopsis spp*	This fungal disease attacks the seedlings, which wilt and die.	Remove and burn any infected seedlings immediately. Spay the remaining seedlings with Benomyl. *As a preventive measure:* If the disease is prevalent, spray routinely with Benomyl, starting from emergence.
Root Rot *Fomes lignosus* and *Ganoderma pseudoferreum*	The tree slowly dies as the roots become rotten.	Dig up and burn all infected trees.
Powdery Mildew *Oidium nephelii*	A white powdery growth spreads over the leaves and fruit, and the fruit drops off before it is ripe.	*As a preventive measure:* Spray with sulphur, Mancozeb, or Benomyl.

Appendix

Measurements and conversions

All the measurements given in this book are in metric units. If you want to convert these to British Imperial units use the values shown below. The Imperial units are the ones in **bold type**.

1 hectare = 10,000 square metres = **2·5 acres**
1 acre = 0·4 hectares = 4,000 square metres
1 square metre = **11 square feet**
1 cm = 10 mm = **0·4 inches**
1 inch = 2·5 cm = 25 mm
1 metre = **1 yard** = **3 feet**

1 tonne = 1,000 kg = **1 ton**
1 kg = 1,000 gm = **2·2 lb (pounds)**
For rough calculations, kilos per hectare and pounds per acre are about the same.
1 litre = **1·75 pints**
1 pint = 0·6 litres
1 gallon = **8 pints** = 4·5 litres

If there is no way of measuring exactly, the following rough measurements can be used:

Weight

A 1 kg tin (of cooking fat or oil for example) will hold about 1 kg of fertilizer. For manure or compost you can use your hands: 1 kg of manure or compost is about as much as you can hold in two hands.

Spoons can be used to measure small quantities of fertilizer. The small spoon shown on p. 9 holds about 6–7 gm of fertilizer when heaped up. The large spoon holds about 20 gm when heaped up.

Length

A tall person's stride measures about 1 metre.

Your hand is about 15–25 cm long. There is a centimetre rule on the cover of this book which you can use to check the length of your own hand. This will be useful in measuring distances between rows when planting.

Volume

A small bucket holds about 9 litres of water.

A 20-litre drum, of the kind used for kerosene, can also be used to measure liquid. A 4-gallon drum (known in East Africa as a *debe*) is roughly the same size. (This type of drum is also useful for measuring manure or compost, and is suggested frequently in the text. If a drum of this kind is not available two small buckets will give about the same quantity.)

Index of chemical compounds

Listed below are the chemicals recommended in this book and the trade names under which they are sold. Different companies use different brand names for products containing the same active ingredient. Farmers should find out which trade products are available in the shops in their locality.

Insecticides

Common chemical name	Trade names	Common chemical name	Trade names	Common chemical name	Trade names
Acephate	ORTHENE	Diazinon	DIAZINON	Endosulphan	THIODAN
Aldrin	ALDRIN ALDINEL ALANDRIN	Dichlorvos	NOGOS NUVAN VAPONA DEDEVAP	Fenitrothion	FENITROTHION FOLITHION SUMITHION AGROTHION NOVATHION
Binapacryl	MOROCIDE	Dicofol	KELTHANE		
Carbaryl	SEVIN CARBARYL	Dieldrin	DIELDRIN KYNADRIN	Fenthion	LEBAYCID BAYTEX
Chlorfenvinphos	BIRLANE SUPONA	Dimethoate	DIMETHOATE ROGOR ROXION	Fenvalerate	SUMICIDIN
Cypermethrin	CYMBUSH RIPCORD			Formothion	ANTHIO
Demeton	METASYSTOX				
Malathion	MALATHION MALADRIX KILPEST	Permethrin	AMBUSH TALLCORD	Profenofos	CURACRON
		Phenthoate	CIDIAL	Protein Hydrolysate	BUMINAL
Methidathion	ULTRACIDE	Pirimephos Methyl	ACTELLIC	Quinalphos	EKALUX
Methomyl	LANNATE			Trichlorphon	DIPTEREX
Monocrotophos	MONOCRON AZODRIN NUVACRON				

Fungicides

Common chemical name	Trade names	Common chemical name	Trade names	Common chemical name	Trade names
Benomyl	BENLATE	Dinocap	KARATHANE	Sulphur (Micronized)	THIOVIT
Captan	KAPTAN	Manacozeb	MANCOZAM DITHANE M45	Thiram	FERNACOL FERNASAN D
Copper Fungicides : Copper Oxychloride	PERECOT CUPRAVIT SHELL COPPER	Maneb	MANZATE DITHANE M22	Triadimefon	BAYLETON
		Methyl Bromide	DOWFUME MC2	Zineb	ZINEB MURPHANE
Copper Sulphate	COPPER SULPHATE				
Cuprous Oxide	PERENOX KOCIDE 101				

Nematicides

Common chemical name	Trade names	Common chemical name	Trade names	Common chemical name	Trade names
Carbofuran	FURADAN	Dichlorpropane	TELONE II	Phenamiphos	NEMACUR
Dazomet	BASAMID	Ethylene Dibromide (EDB)	TERRAFUME DOW EDB 4.5		

Seed dressings

Common chemical name	Trade names	Common chemical name	Trade names	Common chemical name	Trade names
2 Methoxyethyl Mercury Chloride	ARETAN	Captan	KAPTAN	Thiram	FERNASAN D
		Carboxin	VITAVAX		

Herbicides

Common chemical name	Trade names	Safe to use on or near :
Alachlor	LASSO	Maize, beans, sunflower
Ametryne/Atrazine	GESAPAX COMBI	Sugar-cane, sisal
Asulam	ASULOX 40	Sugar-cane
Atrazine	ATRAZINE	Maize
Atrazine with Metolachlor	PRIMAGRAM	Maize
Benzoyl Prop Ethyl	SUPER SUFFIX BW	Wheat
Bromacil	HYVAR X	Sisal, pineapples
Butachlor	MACHETE	Rice
Cyanazine	BLADEX	Maize
Dalapon	DOWPON DALAPON GRAMEVIN	Tea, coffee, sisal, sugar-cane
Difenzoquat	AVENGE	Wheat, barley
Diquat	REGLONE	Tea, coffee, fruit, pyrethrum, sugar-cane
Diuron	DIURON KARMEX	Tea, coffee, bananas, asparagus
EPTC	ERADICANE	Maize, potatoes
Fluorodifen	PREFORAN	Rice
Glyphosate	ROUNDUP	Tea, coffee
loxynil (with mecoprop)	ACTRIL DS	Sugar-cane
Linuron	AFALON	Maize, potatoes
MCPA	MCPA WEED KILLER M AGROXONE	Maize, wheat, sorghum, sugar-cane, rice

Common chemical name	Trade names	Safe to use on or near:
Mecoprop CMPP	METHOXONE CMPP	Wheat
Metolachlor and Metabromuron	GALEX	Potatoes, beans, peas, groundnuts
Nitrofen	TOK	Carrots, brassicae
Paraquat	GRAMOXONE	Tea, coffee, fruits, pyrethrum, sugar-cane
Pendimethalin	STOMP	Maize, pineapples, sisal, sugar-cane
Prometryne and Metolachlor	CODAL	Cotton, sunflower, some vegetables
Propachlor	RAMROD	Onions, sorghum
Propanil	STAM F. 34	Rice
TCA	NATA	Sugar-cane
Terbuthiuron	SPIKE	Sugar-cane
Terbutryn	IGRAN	Wheat
Triralin	TREFLAN	Cotton, beans, sunflower

Glossary

acid	See p. 2
alkaline	See p. 2
Altitude	The height of land above sea level
annual	An annual event is one which occurs every year; an annual plant is one which lives for one year only
bacteria	Bacteria are too small to be seen except through a microscope. Some bacteria cause plant diseases
biennial	A biennial plant grows from seed in the first year and produces flowers and seed in the second year
canopy	'The roof' formed by the branches and leaves of trees
cereal	Plants that have long straight leaves and produce grain, such as maize, wheat, barley, sorghum, millet and rice
certified seed	Seed that is certified as being free of disease by the seed merchant
crop rotation	See p. 9
deficient	Lacking, short of; see also *deficiency*
deficiency	A shortage of something; for example, certain soils may have a potash deficiency, which means that they do not contain enough potash for good crop growth
disease	Diseases cause sickness in human beings; plant diseases make plants weak and unhealthy in a similar way, and they may kill the plants or reduce the yield
disinfectants	Chemicals used to kill bacteria, fungi and viruses on the hands, or on knives and other tools
distorted	Twisted out of shape
dormancy	A period of rest, when a plant or seed does not grow
dressing seed	See p. 5
drought	A long period of unusually dry weather
eelworms	See nematodes
erosion	The loss of soil from the land; soil may be washed away by rain or flooding rivers, or blown away by wind. This occurs when it is not properly protected by a good ground cover of plants, especially on hillsides
fallow land	Arable land, not being used, and left to natural grasses and bush for a season or more
field hygiene	Keeping fields clear of old plant material and weeds; ploughing in, digging in, or burning all crop residues as soon as possible after harvest
fumigation	A special method of killing pests in soil or in stored crops
fungal diseases	Plant diseases caused by things known as *fungi*. These fungi are too small to see, but they produce symptoms such as leaf spots or mildewy growth on the plants. The characteristics of fungal diseases are described on p. 11
fungi	See *fungal diseases*
fungicide	A chemical which kills or stops the growth of fungi, and is therefore used to control fungal diseases
furrow	Where land is ridged by hoeing, or by a plough, the trench between two ridges is known as a furrow
gall	An abnormal swelling or lump, usually produced by a pest or disease
germinate	When seeds start to grow, they are said to 'germinate'
germination	The growth of seeds. Seed has 'poor germination' if many of the seeds do not germinate
herbicide	Chemicals which are used to kill weeds
hermaphrodite	A plant that has both male and female flowers

infest	When insects are found on a plant in large numbers, they are said to 'infest' that plant		known as a 'pollinator'
insecticide	Chemicals that are used to kill insects	pollute	To make something filthy, unhealthy or poisonous
intercropping	Planting two or more crops together in the same field	prevailing wind	In most places, the wind comes from one direction more often than any other; this is the prevailing wind
larvae	The young of insects; caterpillars and maggots are typical larvae	prevalent	Common, likely to occur
laterals	Shoots or branches that come from the sides of the main stem	prone to	Likely to suffer from, or be affected by
		propagation	The process of producing more plants; see *vegetative propagation*
mature	A mature tree is one that is fully grown; a mature fruit or seed is one that is fully ripe	ratoon	A second crop, growing from the roots of a plant, after the first crop has been harvested
mechanically	By means of machines (for example, a tractor and plough)	rhizome	A fat, underground root
		rodents	Rats, mice and similar animals
misshapen	Out of shape, not having a mormal shape	rootstock	In a grafted tree, the plant that supplies the root
mulch	Mulch is material such as grass or leaves, that is laid on the ground to keep in moisture. Black plastic sheets can also be used as a mulch	routine	Something that is done regularly; routine spraying with fungicides is spraying that is done even when there is no sign of the fungal disease
nematicides	Chemicals used to kill nematodes	scion	In a grafted tree, the plant that supplies the top part
nematodes	Tiny thread-like worms; some live in the soil and can damage plant roots	seed dressing	See p. 5
nutrient	A nutrient is something which a plant needs in order to grow, such as nitrogen, phosphate or potash	sensitive	If a plant is said to be 'sensitive' to certain conditions (for example, cold temperatures) it means that it will be damaged or killed if exposed to those conditions
nymph	Immature form of insect		
organic matter	Part of the soil made from rotted leaves, animal manure etc. By adding compost or manure to the soil, its organic matter content is increased	staple food	The starchy food (such as yams, cassava, maize or rice) which people eat most of
perennial	Living for more than one or two years	sterilize	To kill the things which cause disease (bacteria, fungi and viruses)
pH	A measure of how acid or how alkaline the soil is	stimulate	To make, or encourage, something to happen
physiological disease	See p. 11		
pollen	Powdery substance produced by the male parts of flowers; it is needed by the female part so that they can set fruit	subsoil	The hard, rocky soil that lies below the topsoil
		succulent	Juicy
pollinator	Some trees cannot set fruit with their own pollen, but need another tree to supply the pollen for them. This tree is	symptom	A change or abnormality produced by a disease which shows that the disease is present, for example, leaf spots,

	reduced growth or wilting
threshing	Removing the seeds of the crop from the parts that cannot be eaten, for example, separating beans from their pods
tilth	The state of the land. When soil has been dug or ploughed well, and then raked so that it is well broken down, it is said to have a 'fine tilth'
tolerate	To withstand, to put up with
top-dress	To apply fertilizer on the ground, around the base of the plant
topsoil	The top layer of soil, which can be easily dug or ploughed
trellis	A support structure for a plant; it is usually made of wood or wire
truss	A cluster of fruit, for example, tomatoes
tubers	A swollen part of a plant's root, in which starch is stored
varieties	Different types, or strains, of the same crop

vegetative propagation	Growing plants from other plants, without using seed
ventilation	Ventilation is the flow of air through something. For example, if a storage hut has large openings, that are placed so that the wind blows through the hut, then the hut is said to have 'good ventilation'
vigorous	Strong and healthy; fast-growing
viruses	Viruses are too small to be seen but they cause plant diseases
water-logged	When soil is very wet, so that if a hole is dug in the soil a pool of water appears there, then the soil is said to be 'water-logged'
wither	To dry up and wilt
yield	The amount of food produced by a crop (usually in one growing season or one year) is known as the 'yield'